TRY!

Look at Type 2 of civilization!
As we continue; we're destroying life!
Are you able, to decide the fair destiny for your kids' children?

CONSTANTIN DUTCHEVICI

authorHOUSE®

AuthorHouse™
1663 Liberty Drive
Bloomington, IN 47403
www.authorhouse.com
Phone: 1 (800) 839-8640

Published by AuthorHouse 12/12/2019

ISBN: 978-1-7283-3892-7 (sc)
ISBN: 978-1-7283-3891-0 (hc)
ISBN: 978-1-7283-3890-3 (e)

Library of Congress Control Number: 2019920217

Print information available on the last page.

This book is printed on acid-free paper.

CONTENTS

CHAPTER 1

MY FRIENDS

I'm on Mount Athos. I drop my backpack and feel relief. I take off my beret and wipe my forehead. To the south, a fascinating azure sea caresses my eyes—the same magical color east and west. The blue sea holds gently in it giant palm the peninsula of the earth. Huge tectonic forces have raised seabed sediments. The water and the wind dug ditches and deep furrows into the mountains to the sea.

In the south, waves of forests, grass, and stones sweep across the continent. In addition to strength, power, and splendor in nature is the spiritual strength of the Orthodox Christian priests. Through prayers, fasting, faith, and work, they strengthen the steady connection of burning love with divinity. I came seeking for inspiration. Additionally, it is extraordinary that legend says that Mary, the mother of Jesus, and Lazarus, who was resurrected from the dead, set up a place of prayer and refuge on Mount Athos.

I sit on a rock. I reflect.

I came here to find out the truth.

Can we as humans find out where we came from? Where we will go? What the future will be? That would be a more advanced truth. But to be more modest, I would be pleased with gaining at least a partial understanding of what is happening in society. Is the way we are going right? Crimes of all types plague history and today. We need inspiration from Mount Athos.

Why are there so many crimes against nature? Millennial forests, trees, flowers, bushes, and grass are being destroyed. They are living beings!

Birds, insects, and other animals are being sentenced to death. Whales, dolphins, sharks, fish, elephants, antelopes, and felines are hunted without mercy. Why?

I saw a movie in which an elephant was shot in the foot by a poacher who wanted its tusks. With three legs, it could not move. It was constantly screaming. After a week, it fell. It was terrifying. For one week, it moaned all the time, and then it died. It is creepy what torments this natural monument has gone through. What happened to it can happen to all of us. That's awful!

Is there something to be done? First of all, we have to find out the truth.

There are still wars of mass destruction and insecticides, pesticides, fertilizers, hormones, antibiotics, and other chemicals that kill animals and vegetation and threaten human beings. Why?

A hundred years ago, traditional agriculture worked; chemicals were supposed to increase productivity and reduce spending. The poisoning of all nature including people is cynical. All beings—from bacteria to trees, elephants, and people—have an equal right to life—air, water, and food that are not threatened.

If we fight to keep all beings alive, we fight for our lives as well. If we let them be destroyed, it will shortly be our turn to be hunted, poisoned, and destroyed. Something must be done. We must first find out the truth about this scary massacre of nature and investigate alternatives so we can achieve harmony of every living thing on earth. Every creature has to do its part to transform this earth. Terra is a great ship crossing the cosmos like a Noah's ark.

Truth is a burden on my shoulders. I alone cannot do much. I need to talk with friends and share with them my desire for truth; I do not have allies, helpers, buddies, or angels in the sky.

Reader, can you help me? You may also end up helping yourself by working on your enlightenment. With my weak powers, I try to generate new ideas, hypotheses, and directions that you can understand, experience, and broaden. You can agree and develop in this direction, or you can disagree and find more-interesting and effective solutions. I would congratulate you for any success you achieve.

Help can be given through thoughts, words, and deed as those brave priests on Mount Athos say. With thought, you can reflect, meditate, and deepen ideas and change the direction of events in the world as the priests say. In meditation, we set our minds on an object, idea, or activity for as long as we can. If we repeat a text, it becomes a prayer. It is said that a community can meditate on sick people and heal them. So if at the city level or at the country level we all meditated at the same time for ten or twenty minutes, we could change the course of political power. For instance, at noon, we could all meditate for three minutes to diminish the assassination of living beings in nature. Let's try!

We could choose another objective; we are tens of millions—collectively, a huge force of thought, a colossal power pointing toward a constructive goal for the good of all. This would be an interesting experiment. Instead of going into the streets to demonstrate, we would be more united by sharing thoughts of the same goal at the same time change the future. It's worth trying! We can try to heal society by the force of thought.

If you have mercy for the living beings that are being unnecessarily sacrificed, if you have compassion for them, if you feel their pain, their calls for help, you have a soulful communication with Mother Nature. If you realize and reject the injustice, suffering, and pain nature suffers, you have a deep understanding of nature and human cruelty. Meditate and pray. If you feel a sensation as if your heart has stopped, you will wonder if you are a person or an emotion of Mother Nature.

Go to the meat section in a grocery store and consider the pieces of once living animals—pork, red meat, chicken, and more. Imagine their flesh is the flesh of your dear ones. They are touched by incurable diseases. Through misleading advertisements, manufacturers can make you accept that the company recycles them. Horrors! Will you agree you do not have another option.

I exaggerated to scare you a little. But you can see there are similar monstrous actions in progress. In war, people, animals, and vegetation are destroyed. There is a trade in organs taken from the living and dead. People can be exploited just as cattle and so much more in nature is. If we fight for their lives, we are fighting for ours.

I can relate other sinister scenarios based on money, power, corruption, ambition, authority, and so forth. Can you find others?

Maybe you do not have the power to engage in a struggle, a terrible war for life or death. Enemies are powerful, and you do not have the material resources, the soul, or the motivation to fight. You can still meditate and pray; that is a great help. Be concerned for your soul—that is the first step.

By words, you can spread your ideas and those of your friends and connect with others in person, on the internet, on your phone, and so on. That is the second step.

By deed, you can develop associations to preserve the lives of all living beings and unite people of good faith. The concept is to join, form, and organize bands of activists and international fighters to defend life. That is the third step. Some organizations want to support Terra's life including Greenpeace, Save the Trees, Planet B, and Save the Planet. Like them, we want to find alternative and radical solutions to the destruction taking place.

Ideas you develop, compute, experiment with, and use on yourself and your friends are more precious, viable, and practical than prefabricated, imposed, generated, or displayed ideas disseminated by authoritarian powers who do not share your interests.

Authoritarian powers believe individuals do not know the truth and can be manipulated for profit. These authoritarian powers range from those in schools, advertising agencies, and politicians to those in finance, industry, and commerce; there are many.

Your best friends are your parents, those who love you, and natural beings including bacteria, birds, fish, insects, and other animals. Do not neglect them; many can teach you important things. Your best friends also include the inanimate—air, water, earth, stones, and fire.

If you do nothing, others will decide your future. If you act, you can change the future for the good including the harmony of all. Reflect on that.

I sit with my hands on my forehead. I look at my watch. It's five hours until Kostas's boat leaves. It is five kilometers up to Karoulia Monastery past some houses at the edge of the sea. The shortest way is up the west slope for 2,000 meters. I strap on my backpack and beret and go. I'll look for a walking stick because the gravel makes me slip. I do not see any wood around; the trees are down in the valley. There are only rocks, stones, and pebbles. I see a lizard flash by. It stops. I'm immobilized. I look with interest at it and say, "You're a mini dinosaur! I think you're older. You have two lines of white dots on your back. Your paws have five fingers like we do. Breathe and see how your throat moves." Its gray skin blends in with the stones. I say, "I'm leaving." It sprints into its world.

I walk, slip on some moss on the stone I had set my foot on, and steady myself. I have to be careful. The slope down to the valley is steep. I see something moving. A snake hiding behind a boulder. I leave him alone. The road seems dangerous.

Zbrrrrrrr ... A grasshopper lands on my leg. It is about four centimeters long. Strong back legs propel it into the air. Its wings allow it to fly. The hooks on its six legs allow it to grip things. It's a formidable creation. This one has the head of a medieval horse with a guard, large eyes, a mouth with a small palpatory, and long antennae. It can eat flies and caterpillars. I say, "Go," and it goes south like other creatures. Is that a sign for me to go south? I keep heading west.

I see the blue sky, the stony plateau, the sky again, the plateau again, the sky again and again. I slip again and try to stabilize myself, but without support, I roll and roll. Everything is spinning—the blue sky, the boulder plateau, the green bushes. Rolling on the ground hurts my hands, feet, back, chest, abdomen. Everything is in pain.

If I sprawl my arms and my legs, I can stop this Babylon as in judo. I try but cannot spread my arms and legs; it is as if they're paralyzed. Stronger forces than mine are killing me. Every second seems to be an end-of-life eternity. I feel a pressure on my abdomen that stops my rolling. I open my eyes. Everything is spinning—the sky, the plateau, the bushes. I hear someone but cannot understand what he is saying. I say, "I'm ... hurt ... Help me. Who are you?"

After a while, I focus on someone's silhouette. A monk in a cloak with a stick in his hand is staring at me. He has a foot on my belly. I say, "Get your foot off me! It hurts!"

He bends down and turns my head to the hill. He says, "This is better. Relax. I will try to alleviate your pain by massaging you. Do not move." He massages me up and down my body. I feel the pain lessen.

I ask, "Who are you?"

"Be quiet. You do not have fractures, only blows. They hurt, but the pain will go. My name is Eohn. I'm a monk. And you?"

"My name is Madah. I'm an engineer. I have come to find out the truth."

"You chose a very hard way to achieve that. It's a painful way. You can feel that. You have to go through trials. It is not known whether you can overcome them. The pain urges you to quit. You can take that as a sign from above. Many broke their teeth on the truth. It is hard to achieve even a little understanding. Many illustrious people who have tried to find the truth have advanced just a little. You with your little knowledge want to understand and explain the history of millennia, to decipher the deeds of men known and unknown. That is hidden, guarded. You know you can lose your life for revealing mysteries that do not convince the superior powers."

"I will take the risk," I say. "I would not be the first to sacrifice himself. Many were eliminated, exiled, tortured, imprisoned. We do not know them all. History repeats itself. But I cannot succeed alone. Can you help me?"

"Of course. I can help you with all my heart, love, and understanding. I will help you with everything I know. I'm older, as and you know, changes come to me differently than they do to you young people, but what the cyclical movement is, has been, and will be is valid for people as it is for nature. I know other things you do not see. You can imagine that you are me. When you need me, think to me, call me, and I will come to you. Understand?"

I nod. I understand.

"Let's go," he says. "I'll take your backpack. Lean on me and we'll get to Karoulia."

How did he know my destination? I wonder. I get up, and step by step, we get to the shore. It hurts, but I resist my pain. I am in no mood to admire the wonders around me. He finds Kostas and tells him to get me to my plane.

On his boat, the *Athena Thessalonika*, we sail for ten hours. I want to admire the Aegean Sea, but I cannot. I want only my mother. I sit on my bench. I close my eyes. After a while, someone sits next to me. I open my eyes and see a female leg. I close my eyes. After a while, I look at her. She is reading something. A magazine. I see she is reading an article about UFOs, testimonials, radar confirmations, and so on. I say I'm interested in that as well. I look at her face. She is a goddess. She harmonizes in her entirety. She seems to have an athletic build. I cannot take my eyes off her. She turns her gaze to me. I have never seen such eyes; their blue-green color radiates power. Her energy passes through me as an emotion from my eye through my brain through the back of my spine—my hands, feet, body experience tingling in waves.

She has extraordinary hypnotic powers. Even my pains are diminished. This goddess instills in me a sense of mystical adoration—admiration for her beauty and respect and fear of her powers. She is a fascinating, harmonious creation, a great pleasure to gaze on. I imagine her as a statue at an altar to which I pray when I need help.

"Do you hurt?" she asks.

I nod.

"I do massage to alleviate pain. Look into my eyes and relax, relax. Breathe deeply. Breathe deeply. You pain is leaving you."

Her hands move from my head, to my chest, to my abdomen. She repeats that a few times. A stream of heat comes from her hands, and thrill and excitement come from her eyes.

"Is it better?"

"Yes, thank you."

"Don't mention it."

She goes back to her magazine. I do not know how to get in touch with her. When I look at her, my inspiration disappears. "Are you interested in UFOs?" I ask.

"I want to see what they're all about. They're stories written for money, to satisfy the curiosity of the naive. UFOs are presented as Wild West bandits. I'm exaggerating a little."

"My name is Madah Lulac. I'm very interested in UFOs. I've developed some ideas about them. I think there's a high probability that other civilizations exist."

"My name is Ahve. I'm convinced that there is life elsewhere in the universe, that there are other intelligent beings. They bring life from one planet to others. They heal. They do not kill."

"Do you say we humans are destroying nature?" I ask.

"Yes. But you better sleep, maybe thirty to sixty minutes to reinforce the restoration."

She moves her hand in front of me saying, "Breathe deeply. Tell yourself to go to sleep."

And I fall asleep.

After a while, I wake up. I feel better. "Ahve, I was on Mount Athos. The activities of the monks are impressive. Can you get my book out of the bag? Find and take out the New Testament where I have a prayer sheet as a bookmark ... Thank you ... Here you go. These brave monks keep old traditions. They connect with the universe. They persevere in maintaining a harmonious connection between the divinity and people based on rituals, faith, and love. They are rewarded from time to time through revelations and miracles. They are just, fair, and honest. They help those in need, the sick, the hungry. They are an excellent example for people. Religion contradicts science. It is different from this. Through science, we have theories, calculations, and technology—planes, cars, cell phones, TV, music, movies, nuclear power stations, and much more. Religion cannot give us those."

"That means we do not need them through religion," she says. "Science can explain much of nature; it generates technologies that can solve problems. That's the good part. The sad, damning part is how science and technology are used. Today, it is used as a weapon to battle all nature—plants, animals, and people. These two paths should be complementary.

"Compassion in religion could transfer to technology to benefit living beings. The universe is infinite, and the ways of knowledge are many—science, religion, contemplation, meditation, and others. In their way, all contribute to the advancement of humankind. If we take only one example—science—and eliminate the others, it becomes an oppressive dictatorship that generates injustice, abuse, and crime. There have been such cases in history, and there are cases of the same today."

CHAPTER 2

DELIBERATION

I went home almost dreaming. I do not remember much. Eohn and Ahve had been compassionate with me; they made sure I got back. They told me to sleep as much as possible.

Due to their magical powers, I woke up a day and a night later. When I woke up, I felt a great affection for them and a terrible longing to see them again. I called them, and they finally returned my calls. I told them I wanted to see them again, and they agreed to that.

We meet at my place. I prepared some food, got wine, beer. I'm waiting. The first to come is Eohn. Ahve also appears. I ask what they want to eat. They say they do want anything. I propose beer, wine, coffee. They refuse. "What can I give you?" They want tea. I look for and find black tea without sugar or milk.

"I eat only when I'm hungry," Ahve says. "I'll ask you if I need something, or I'll just get it. Let's talk about what you want to do."

"Same with me," Eohn says.

I bring paper and pencils, and we start.

Ahve says, "There are four chairs. We can move one away."

"This chair is for Reader," I say. "He has opinions. We include him in the discussion, agreed?"

They nod.

"I want a beer to start," Reader says.

I bring him a beer and a glass.

"What do we start with?" Ahve asks. "With the creation of the world? The earth? The universe?"

I say, "No. The creation of the universe is unique and is a difficult phenomenon to study. I turn my attention to cyclical events. They're easier to observe, analyze, and compare. The succession of day and night ... The cycle is endless ..."

"But today is not the same as yesterday," Reader says. "One day is longer or shorter than others, right?"

"Correct," I say. "Besides the cyclical repetition, there is a change. The change is small, but in the long term, we come from summer to winter and back to summer."

"The heart beats cyclically," Ahve says. "It beats more frequently in children and can be disordered in old age. A second is close to the duration of a heart cycle. Breathing is cyclic. The functioning of human organisms is mainly cyclical. We also find in nature an annual period of vegetation and animals."

"History is full of changes," Eohn says. "One kingdom falls and another rises. Think of the Egyptian, Roman, and Mongol empires for instance."

"Today's great powers are the United States and Western Europe. Do you see an end for them?" Reader asks.

"Of course they may go down," Eohn says. "Greed—having as much as possible—and the desire for domination and control is destroying life—forests, grass, animals, fish, birds, bacteria—everything. People will have less and less until they die. A terrible weapon in this battle is money—financial politics. With the help of money, they lure and deceive people to be destructive."

"We have a chance to avoid this destruction," I say. "We have the opportunity to develop peaceful paths to meet the needs of people and beings on Terra. We can spread life into the cosmos. That's why I need your help. Do you agree?" Everyone agrees. "Life is cyclical and changing. What happens today happens tomorrow with small differences. Change constitutes life in the long run and makes life more and more difficult for living beings. I want to develop a practical way to harmonize life on Terra.

"I divide civilization into two stages. The first is civilization to date. It has accumulated good, useful things—science, technology, arts, and religion. It has also accumulated destructive, damaging things. The second is the civilization of the future. We are expanding human activity and living beings in our solar system. We develop ecosystems that work harmoniously among humans, animals, and plants."

"Nikolai Kardashev has classified civilizations this way," Reader says. "Type one uses resources on his planet until they are exhausted. Type two uses all resources in the solar system, and type three uses all the resources in the galaxy."

"We're making a change," I say. "In type two, they do not deplete the energy resources of their solar system but increase them. These resources will be sufficient for a very long time. We are dealing with this."

"It's interesting to multiply energy," Ahve says.

"It's hard to believe that's possible," Eohn says.

"We're advancing new knowledge about energy," I say. "We'll see that the concept of energy is very elastic—we can have as much as we want. We can generate energy or we can consume it."

"Generating energy from nothing is not possible," Reader says. "Producing it is exciting, but what does it consume?"

"By developing new knowledge about energy, we can generate it," I say. "We can use the same energy several times, and we can do so more efficiently. Out of the surrounding area, an artificial construct can extract energy. If we have excess, we'll want to consume it."

"I'm not convinced," Reader says. "Give an example of how we could use energy several times."

"We hit a nail with a hammer," I say. "We could attach a spring to the hammer. When it hits a nail, it returns close to where it was. We continue until the hammer's energy is exhausted. We can do the same thing with the same energy several times. It is a simple example of multiple uses of the same energy."

"We can come up with an invention," Ahve says.

"Those machines that pound pillars into the ground work similarly," I say.

"I failed this invention!" she says. "We expect others. Let's go further. Tell us how we get to the type-two civilization."

The type-two civilization is different from what's on earth today," I say. "The possibilities are enormous. We have accomplishments that seem fantastic, dreamy, incredible. They will amaze you. You can't imagine how wonderful the future could be—another world! We have a rare chance to advance civilization. If it is lost, it is lost forever."

"Tell us how we get can get there," Reader says. "It could be fantastic. For us, there is great hope. If it is true, we can try our best to get there."

"To reach the type-two civilization," I say, "we need four things.

First, rapid-transport systems with constant acceleration and deceleration and low energy consumption. It can operate in the cosmos as well as on Terra.

Second, we need power generators for space and terrestrial applications.

Third, we have to depollute—reduce waste to the atomic elements of which it is composed.

Fourth, we have to treat all life ethically, and that includes even microorganisms.

I'll make a brief presentation on these points. Later, we'll move on to the theory that allows us to achieve these goals. There will be examples of how the problems are solved today as opposed to solving them in type-two civilizations.

"Four people can drive two hundred kilometers at a hundred kilometers per hour using twenty liters of fuel.

With type-two knowledge, we could build a vehicle that does the same thing but actually produces energy, the equivalent of twenty-two liters. Airplanes and ships could do this as well."

"Impossible!" Reader says. "Perpetual motion machines have never been built. No one has ever produced energy from nothing. Are you an engineer? How can you sustain such utopias?"

"I think it's impossible," Ahve says.

"God is great!" Eohn says. "Maybe there are miracles people can perform."

"Remember your emotional states of fear, distrust, wonder, and more." I say. "These are caused by your shock on contact with type-two civilization. Certain events are far from our understanding. I suppose contact with them should be progressive, and that's the purpose of our discussion.

"We are building another type of vehicle that moves with constant acceleration and decelerations. For a = 10m/s^2, the time A to B is 282 seconds, about five minutes, twenty-five times shorter than what current machines can do. Fuel consumption is 0.8 liters, twenty-five times less than currently. We can reach the moon in about 3.3 hours and Mars in about 35 hours. Transport is fast, and it allows us to travel through the solar system quickly. Transport of the future is the support base for the next development of humanity.

"In addition, we can assess technological advancements. If we multiply, we shorten transport times by a factor of twenty-five, and we can lessen fuel consumption by a factor of twenty-five as well. Twenty-five times twenty-five is 625. Type-two transport will give an advance about 1,000 times greater than current transport."

"It's a miracle, incredible," Reader says. "But I don't want to be deluded."

"It's too good to be true," Ahve says.

"I hope to understand this, and I hope others will want to go in this direction," Eohn says.

"The second element is power generators for space and terrestrial applications," I say. "Some machines extract energy from the environment. We can have power sources from 10kW to 1000kW. This is a convenient range for applications."

"It's a smaller big bang," Reader says.

"It can be similar," I say. "Perhaps through this, we can advance understanding of the big bang."

"That's not how we're going to get a lack of energy," Ahve says. "I understand that it is no longer a depletion of resources. We use how much we need, and the rest remains."

"That's optimistic," Eohn says. "We hope it will also be useful for humans."

"Third," I say, "we depollute by reducing waste to its basic elements. We start with the idea that we have inexhaustible sources of energy and use that in the depollution process. With energy at high temperatures—in a 3000gr to 1000gr C thermal reactor—we can dissociate waste and bring it to form its basic chemical elements. It's a form of radical depollution. These chemical reactors need high-fusion metals. Metallurgy will develop in space, where we escape gravitational forces. Other classic methods remain to be used and developed—sedimentation, filtration, centrifugation, fermentation, and so on."

"We can depollution the whole earth," Eohn says, "and in the future, we give up polluting activities for the good of all living things."

"Depollution will be the technique we use in space and on land," Ahve says. "Waste will be eliminated."

"Do you see this as a religion, philosophy, or something else?" Eohn asks.

"I see it more than philosophy and less than religion," I say. "It is a belief. It can migrate to religion if we rely on miracles as you saw above. But they fall into the scientific category; they can be explained and used. I think the concept can lead to power and spiritual development such as distance communication and others. This vision allows the invention of new symmetrical and mental exercises that would be a new way in the field of harmonious human development."

"If what you said was true," Reader says, "that would be a platform for future political policy that would be of great interest to people."

"It's possible," I say. "Humankind wants an attractive, harmonious, and painless future."

"You present an idea of a civilization that is efficient and harmonious," Ahve says. "It can keep and increases its resources for the benefit of all—heaven on earth!"

"Presently," Reader says, "people are exhausting our resources and destroying other living creatures. That is making people's lives more difficult—a hell on earth. That cannot last much longer. We must do something to change current concepts and find new solutions, other ways of advancing civilization even if they are unusual."

"When we have an abundance of resources, space, and energy," Eohn says, "that will make conflict no longer necessary. Cooperation and harmony with nature will prevail, and we will no longer need to exterminate life."

ALIEN VEHICLES

"I have three theories about UFOs," Ahve says.

"One, they are inventions, fantasies for the press to attract the public's attention and make money. People seek them, accept them, and hope for a change for the better.

Two, they actually exist. They have extraordinary performance: accelerations—0 to 100000m/s^2—fabulous speeds that get them to other solar systems and are driven by intelligent beings who interact with matter and living beings via telepathy. They're not aggressive.

"And three, they are atmospheric phenomena that can be explained. A small fraction of unexplained remains can be attributed to more-evolved civilizations. We are not alone in the universe."

"I agree," I say. "Inspired by these, we can imitate them and see what they are showing us. I believe they are inviting us to decipher and approach them, but direct contact would be detrimental to us. Approaching will be progressive, lengthy; it's a test for humankind. Ethics is the biggest obstacle. People have abjectly destroyed living beings as they please. Other civilizations carry life from planet to planet, spread it, cultivate it, and preserve it."

"I have heard of cases of kidnappings of people, animals, property destruction, UFO crashes, and others," Reader says. "How is that?"

"We do not know. We have to study that," I say. "First of all, we will look at the transport and energy issues that are the foundation of superior advancement. They lead to further extraordinary developments. We need to remember that ethics and depollution are extremely important. They are mandatory. Without them, we will not advance. Once we understand and use them, we can approach being a type-two civilization."

"Can we with our current level of knowledge understand higher levels of knowledge?" Ahve asks.

"I think it's partly possible," I say. "Space vehicles, UFOs, are also subject to physical laws; they are moving through them. They know more laws that we do. We are slowly advancing toward that knowledge. The results can be extraordinary as presented above."

"UFOs are supposed to have an accelerations of 0 to 100000m/s^2 and even greater," Reader says. "People cannot survive such forces; they'd be flattened!"

"Let's consider that problem in two steps," I say. "One, acceleration applies a force to a mass. The force is reduced to the center of the mass, and we say we have a force applied to a point, the point force. This is the case with the forces that people use on earth.

"Two. Another type of mass force applies to every point of the body—the vehicle. It does not reduce to the mass center. It accelerates the vehicle as a whole; we do not feel it. There can be accelerations as great as possible. This is the case with UFOs. Consider this. A man in free fall is accelerating without feeling anything. All his constituent elements have the same acceleration except for friction with the atmosphere.

"A satellite is attracted to the gravitational force but is subject to centrifugal force. Cosmonauts do not feel these forces. I say that UFOs use this kind of effect. They can do stunning maneuvers in this way with no damage. Punctual force applies to the center of mass; the vehicle is accelerating, and people feel powerful forces that affect them. In this case, no higher accelerations can be applied, perhaps 1g to 3g (10 … 30m/s^2); they are modest compared to a UFO. Is it clearer?"

The others say, "A little, but we think there is something else."

"Yes there is," I say. "They can generate acceleration, propulsion forces through their own resources, an idea that is not tolerated by current human knowledge. Here is a leap in our knowledge. We come to show that it is possible. That tells us they do not need gas propulsion as rockets do. They preserve their original mass throughout space travel. They are moving in accelerating motion. They can attain very high speeds; they travel through interstellar distances in a short time: months, weeks, even just days.

"The space vehicles that visit us are from solar systems close to us, just tens or hundreds of light-years away. These vehicles have performance unequaled by their level of evolution. As you know, some are disk shaped while others are cylindrical or spherical, and they have different performance."

"But the most advanced can give their knowledge to those who are weaker, devoid of knowledge!" Reader says.

"No, it's not like that," I say. "Each civilization has its own level of development. They do the best they can. If they received more-advanced knowledge, they would not value it. It is not to their advantage to distort their evolution. Every civilization is building its own way. The presence of UFOs on Terra is a sign that we are below the level of our neighbors' development and are on the wrong path. We have to change human activities for the good of all terrestrial creatures. They do everything they can, and we have to do the same."

"Do you have a more advanced view?" Reader asks.

"Yes," I say. "It is far beyond the current human understanding. It's from the emergence of life on earth, 3.5 billion years ago, with bacteria and later plants. God created them. Another hypothesis is that life evolved from inorganic matter. I think life came from other galaxies brought here by highly developed civilizations. They spread life throughout our galaxy. Different developments have taken place in the solar systems close to us. Some humanoids are far ahead of us on the path of evolution.

"These super civilizations have vehicles far above UFOs' capabilities. They travel between galaxies. Their knowledge goes beyond human understanding. I think I can create matter, energy, life. Maybe they are like gods. I have not arrived at the point that I can understand such super civilizations. They have been ahead of us for over 4 billion years! We think that different races on earth—white, yellow, and black—are found in solar systems near us. They came to earth and stayed here. It's a hypothesis."

CALCULATIONS RELATED TO MASS FORCES AND ENERGY.

My friends went home, but we met again in a few days.

"I liked math and physics," Ahve says, "and I look forward to seeing what you can show us."

"I was good at math, but I hope to keep up with you," Eohn says.

"I never shined scientifically, but I promise to study what you say," Reader says.

"I made tens of thousands of calculations to come up with a valid solution," I say. "Every calculation is due to an experiment. At first, they did not yield positive results, but each one was a lesson about the path to be followed. Keep these in mind until you understand them."

"Reading them isn't enough?" Reader asks.

"If you repeat a calculation, you will understand it better," I say. "Understanding is personal—different from person to person. Answer this—how do we generate force?"

"We accelerate or decelerate a mass; it opposes inertia," Ahve says. "So we obtain a reaction force. It has the value resulting from the acceleration product and its mass.

F = a * m; [N]

"The result is in N or Newton. For: a = 1 m/s^2, m = 1 kg

F = 1 * 1 = 1 N

"I remember from physics that the distance between A and B is measured in meters," Reader says. "It has submultiples—centimeters or millimeters—or multiples—kilometers. Speed is measured in meters per second, m/s. That shows us how far we went in one second.

Acceleration is measured in meters divided by the square of the second, m/s^2; that shows us the rate of increase or decrease of speed in one second."

"I say that the mass is measured in kilograms," Eohn says. "In the old style, the mass of water with the volume of one liter or of one the decimeter cube is one kilogram. It has submultiples, gr for grams; multiples—one tons equals a thousand kilograms and others. Time is a fraction of the earth day—$1/86400^{th}$. It's close to the time between two pulses of the heart."

"Very good," I say. "You understand the basic units. Everyone can use them. The advance of physics has allowed us to define these sizes with great precision. The most accurate clocks are atomic clocks, which are accurate down to seconds over many millions of years. For distance, one meter is now defined as 9,192,631,770 oscillations of a cesium-133 atom. As you can see, it is difficult for the reader to try to reproduce the units of measure in the modern vision. We remain at the simplest ones formulated before.

"We start with two basic ideas. One, we want to get a mechanical force by accelerating a mass. That is the current method used by rockets, planes, and ships. Once accelerated masses are expelled, they leave the vehicle. A rocket is capable of accelerating as long as it has fuel, fifteen to thirty minutes. Next, it travels through space based on the speed attained. The limited propulsion time limits the use of these vehicles drastically. We want to have a much higher function—hours, days, months.

"Second, an accelerating or decelerating mass generates mechanical forces; we keep the mass in the propulsion system; we do not lose it. Complex translation and rotation movements give greater impulses in the forward direction; it gives smaller impulses in the opposite direction. In this way, we get a propulsion force that moves the accelerated vehicle. We travel great distances in a short time—to the moon in three hours and to Mars in thirty-six hours."

"But it isn't possible to obtain propulsion in an isolated physical system in space," Reader says. "Nobody has done that. The laws of action and reaction, energy conservation, and others are being violated."

"That's right," I say. "Today, we do not accept the idea that we can generate a force, an impulse in an isolated vehicle. But people have in the study EmDrive, a cone in which with RF oscillations 2 GHz to 9GHz generates a force, as a small value of mN. There is also UFO propulsion that is a model for us. So it's possible. We are going to do the work that will bring us closer to other civilizations—a great step for humankind and a very small one for the universe."

"We should do everything to advance this work," Ahve says. "We can get a positive result. It will be good for all beings on earth."

"Let us recall a few notions of mechanics," I say. "Figure 2.1 uses it to show how we produce a force. It puts a man on roller skates in motion. A hand pushes with a force

F = 100N, called action. Mass m1 = 10 kg is accelerated with a2 = 10m/s^2. It opposes a counternegative force, -F, the reaction force. Always, the algebraic amount of action and reaction is null. It is universal law valid all the time.

F-F = 0

"If they are equal, how can we propel anything in space?" Reader asks.

"It is possible in more-complex conditions to get forces to move vehicles," I say. "We add a reduction in consumption by energy recovery, qualities that offer qualities out of the ordinary far exceed all current achievements.

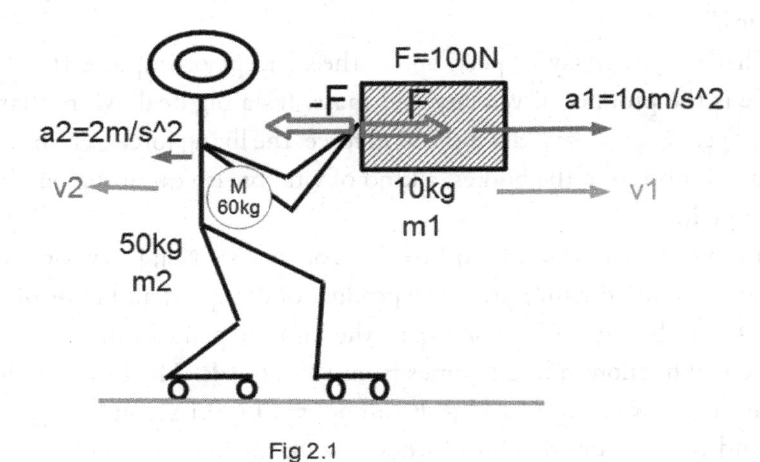

Fig 2.1

"Return to figure 2.1. We see that acceleration

$a1 = F/m1 = 100N/10kg = 10m/s^2$

"Same,

$a2 = -F/m2 = -100/50 = -2m/s^2$

"As we can see, we did this experiment to get the -F force that moves the man on wheels. Instead of m1, you can imagine the rocket gas, the air exhausted by the engine of an airplane, or the water pushed by the ship's propellers. In these cases, m1 is evacuated, disappeared; we do not want this mass loss.

"There are also elements that are not immediately visible in this experiment. They are hidden, somewhat mysterious. They belong in fairy tales. We have to use these.

"Speed is obtained from the acceleration product and drive time.

$v = a * t$

"After one second,

$v1 = 10 * 1 = 10$ m/s; $v2 = 2 * 1 = 2$ m/s

"After two seconds,

$v1 = 10 * 2 = 20$ m/s; $v2 = 2 * 2 = 4$ m/s

"After three seconds,

$v1 = 10 * 3 = 30$ m/s; $v2 = 2 * 3 = 6$m/s

"You can continue as an exercise."

"But speed is not hidden," Readers says. "It is displayed on cars, planes, and ships."

"That's right," I say. "Speed enters the concept of impulse (P). The impulse is the product of velocity and mass. It's a big deal. More than a hundred years ago, it was called the vital force, the living force that showed the state of motion of the bodies, a kind of vital breath of movement. Try to imagine it.

$P\ m * v = m * (a * t) = (m * a) * t = F * t$, or [N * s = Ns]; [Newton * sec]

"We read that the impulse is the product of the speed and mass of the body. Or another form: F * t shows us the impulse gained during t given by force F. The more subtle t comes from t2 - t1 = dt. The dt is the time interval in that we measure. This dt can be small—0.1 second, 1 second, 10 seconds, etc. Under this form, we will work further.

"After one second,

$P1 = 10 * 1 * 10 = 100$Ns; $P2 = 2 * 1 * 50 = 100$ Ns

"After two seconds,

P1 = 10 * 2 * 10 = 200Ns; P2 = 2 * 2 * 50 = 200 Ns

"After three seconds,

P1 = 10 * 3 * 10 = 300Ns; P2 = 2 * 3 * 50 = 300 Ns

"It seems that at equal speeds, impulses are equal," Reader says.

"Equality comes from equality of action and reaction," Ahve says.

"That's right," I say. "In the future, we will process these impulses to get the desired movement of the vehicle. It is remarkable in our case that the common mass center M between m1 and m2 is fixed. It retains its original position. It is a reference for measurements and calculations; it is virtual, so we do not see it. For a vehicle, airplane, or rocket, there is a mass center common to the masses of motion fluids; what is immobile stays in the same place. A rocket leaves the launch pad, but the virtual mass center—the mass between the vehicle and the exhaust gases—remains unchanged. The virtual mass center cannot move in our understanding, but type-two civilizations can do it."

"It's weird to see planes, to see movement, but not see a fixed point," Reader says. "It's hiding."

"I think so too," Ahve says. "We do not see the gas expelled. The common mass center is virtual, not seen. It is not in our head to see the whole phenomenon; we have not been taught. We usually see part of the events, the side that is popularized, published, which suits those who present it."

"Here's an example," I say. "The sun and earth have a mass center close to the sun. The sun is spinning around this virtual point in a small orbit while the earth is moving around it in a much larger orbit. This point is very stable regardless of the oscillations of the two bodies. People cannot change the position of the mass center. Of course, a type-two civilization can do it. We will also reach this possibility."

"This displacement is very dangerous," Eohn says. "We see the world damage the natural balance of living beings and inanimate things. If you say we have a way to make these changes, is it not dangerous?"

"Yes, it is very dangerous," I say. "UFOs in the hands of unscrupulous people will destroy the planet shortly. That's why we stood in the balance as to whether to reveal it or not. The earth is headed toward catastrophe. Life is being destroyed. Powerful countries exploit weaker countries. But with very advanced means, rulers and populations can turn from destruction and choose the path of harmony. Humankind has alternatives."

"You have chosen an altruistic path," Ahve says. "You want to do good for humanity. We will help you according to our abilities. I hope humankind will not be destroyed, but it has to awaken. I believe there are yet people of integrity who can help us."

"There is another important consideration—energy," I say. It is calculated with the formula

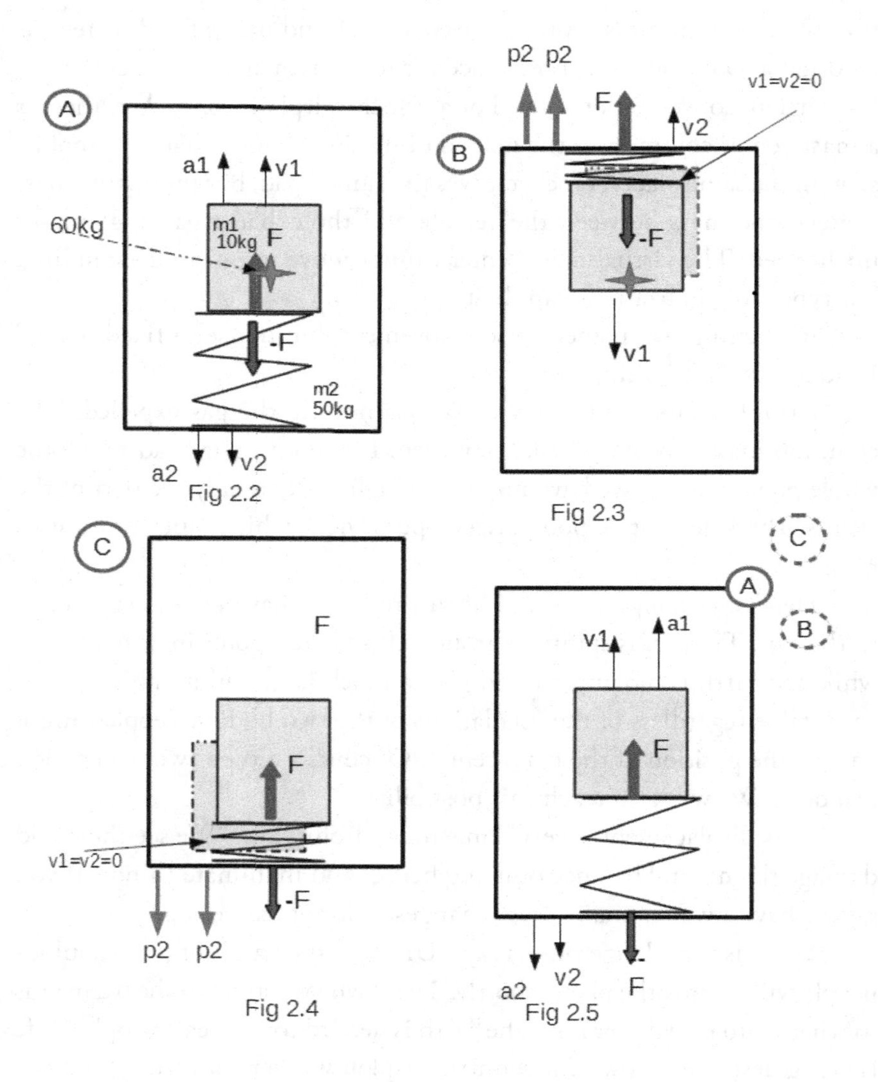

$E = m * v^2/2$; where m is the mass of the body and v is the speed of a point from the origin to us from the 60kg mass center. It is measured in Joules or J.

"After one second,

$E1 = 10 * 10^2/2 = 500J$; $E2 = 50 * 2^2/2 = 100J$

"After two seconds,

$E1 = 10 * 20^2/2 = 2000J$; $E2 = 50 * 4^2/2 = 400J$

"After three seconds,

$E1 = 10 * 30^2/2 = 4500J$; $E2 = 50 * 6^2/2 = 900J$

"The small mass receives more energy," Ahve says. "We want it to be the other way around—the big mass should be able to move faster. If we choose another reference point, the masses in motion will have another distribution of energy. We would get erroneous results; we would have illusions of the impossible. You can try."

"As you see," I say, "the masses m1 and m2 get same impulses, the result of action and reaction, F and -F. They have different speeds and different energies. The lesser mass receives more energy and the larger mass less energy. Next, we experiment with a closed box."

"In this case, we have a back-and-forth movement of a box, oscillation," Ahve says. "The center of gravity is immobile."

"True," I say, "but we will change our system to study energy recovery."

"This change is imaginary," she says. "It helps us understand the multiple reuses of the same energy. It is a big step toward substantial energy savings."

The Case of Linear Oscillatory Motion

"Consider a box in space," I say, "or on wheels on the ground, or a boat on water. Inside, we have a mass, m1, that is moved by force, F. The symbol for the force is a spring that pushes m2 and m1. Consider F the force of action and the -F reaction force. On the wall, a spring pushes back the mass m1 in the opposite direction at the same reverse speed."

"There is no perfect elastic repulsion like $v1 - v1 = 0$," Reader says. "Springs can restore an average of eighty percent of the energy. People do not have such perfect things."

"That's right," I say. "I want to simplify things, not get into the details. Eliminate the elements of friction, the loss, so the attention is directed to the essence of the phenomenon.

"In figure 2.2, we have two bodies of mass, m1 and m2. We have two opposing forces, action F and reaction -F. We focus on vehicle movement m2. We want to see what happens to it. The spring pushes m1 with force F and m2 with -F. They receive impulses opposite but equal in value. A is the starting point, and m2 moves to B.

"In figure 2.3, m2 reaches point B and v2 = 0, m1 has the dotted line drawing, stops motion, v1 = 0. It receives a pulse p2 = m2 * v2 = m1 * v1. It is the stopping state that has a short duration. Further, the spring moves the two masses. Mass m2 receives a pulse p2 = m1 * m2 * v2 = v1. It moves to point C. Note that mass m2 receives twice the impulse, p2, once it stops the movement. The second time it moves it in the opposite direction. It is remarkable that the center of mass, weight m = 60kg, is immobile The mass assembly, m1, m2, remains in the same place.

"Figure 2.4 repeats the previous movement but in the opposite direction. Mass m2 reaches point C. It gets p2 and stops along with m1; it's shortly immobile. It receives another impulse p2 and moves to point A. So the mass m2 receives twice the impulse p2. Once it stops the motion, the second time, it runs in the opposite direction.

"Figure 2.5 shows the oscillatory motion cycle. Our box moved back and forth from A to B to C and back to B. The magnitude of the oscillation is the BC distance. Without friction, oscillation takes a long time."

"Why consider swinging boxes?" Reader asks. "Can that propel UFOs?"

"No," I say. "The movement is simple, and it lacks the propulsion we are looking for. First of all, it is an experiment for you. Try to redo it practically. Take a box on wheels, rollers, pencils, arc, rubber, and make a mass oscillate inside. We will find familiar elements with the mechanisms we want to do next. Practice! It is essential to have insight into the impulses, forces, speeds, accelerations, mass center, and more."

"I see that the impulses at the end of the race are important," Ahve says. "In B, p2 = m2 * v2 + m2 * v2 = 2 * m2 * v2. An impulse stops the movement, and the second gives a movement in the opposite direction. I feel that with return speeds, it is essential that the spring sends with the same speed, v2, at mass 2, is a hundred percent. People still cannot do this, but nature can. Springs have eighty percent efficiency. They recover energy, but more-efficient procedures have to be found. This reuse of energy will be one of the important concerns in the future, a measure of progress."

"It must be remembered that the mass center remains immobile," Eohn says. "This is due to impulses in B that are equal to those in C. To get a movement, there should be an imbalance."

"We're going further," I say. "We're making modifications to the mobile vehicle, the swinging box. These changes in a simple case give you an idea of inertial propulsion. The model does not exist in practical applications. This virtual model uses it to evaluate energy, power consumed for propulsion forces.

The starting point is in figure 2.6. Our system is moving symmetrically. I put the impulses given by m1 at the top. On the bottom are equal but opposite impulses. The mass center remains immobile. There is only one oscillation between the two limits of the box.

"We are making a change that suggests future propulsion systems. Figure 2.7 shows that mass m1 produces a pulse p1 = 2 * m1 * v1. At the bottom, less: p1 = -m1 * v1. This is a model that comes from more-complex mechanisms. With this, we do some essential studies on energy consumption. Note that mass m1 has on top v1, and on the bottom reaches a half speed of 0.5 * v."

"That's impossible!" Reader says. "The laws of physics are not respected here!"

"This is a model for study," I say. "The speed v1 cannot reach half the value in figure 2.6. Figure 2.7 is a different model. I wrote 'virtual'; it does not come from the previous figure. These values are coming from future studies."

"Then why do we not study the more complex ones?" Reader asks.

"It's easier to introduce new concepts based on the simplest models," I say. "It's simpler for others to understand them that way."

"So the models belong to other future studies," Ahve says. "The drawings shown do not derive from the oscillating moving box. Let's see what's coming up."

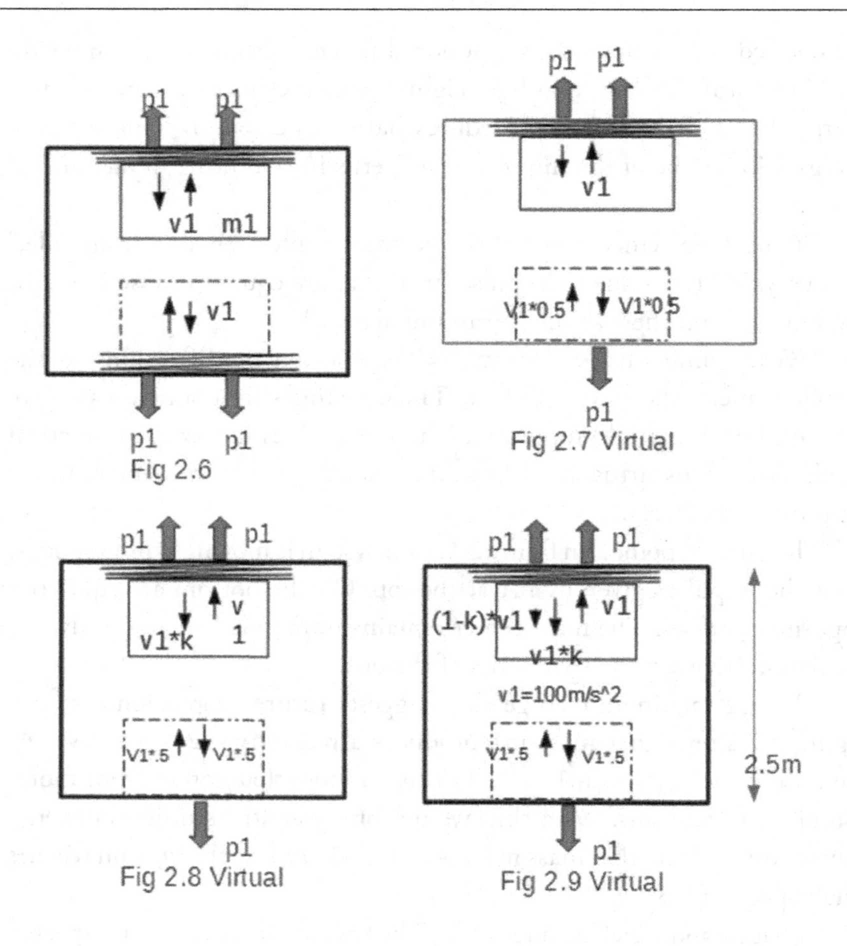

Fig 2.6

Fig 2.7 Virtual

Fig 2.8 Virtual

Fig 2.9 Virtual

"Figure 2.7 shows that the box is moved by 2 * p1 upward, down by a single pulse p1," I say. "There is a difference.

2 * p1 - p1 = p1.

"This impulse gives upward movement. It is a very important observation that the mass center goes upward—it is moving. Each cycle repeatedly changed position. The box moves, accelerates cyclically, and acquires a pulse p1 at time intervals, T."

"This common mass center is immobile in human practice and experience," Ahve says. "If we consider the vehicle and the masses that push them, it stays fixed. I see the danger in this. We can change the mass center between the sun and earth, but we might cause an imbalance in the entire planetary system."

"That danger is very high," I say. "That's why we have to proceed with caution and have a solid, humanitarian ethic."

"I see this movement as provoked by external forces such as God," Eohn says. "External forces can intervene everywhere. It's one way to look at it."

"I'm not going that far," I say. "We have to develop applications that belong to people. This movement of the center of mass remains essential. We have to take this into account. Let's move on to make discoveries about energy. These are extraordinary.

"Figure 2.7 shows how the mass m1 has the speed v1 upward. Speed v1 is downhill with the same value, v1. In practice, people cannot achieve this equality between the speed of incidence and the rate of return, but nature can.

"In figure 2.8, v1 has the return velocity v1 * k where k = 0 … 1. This means that for k = 0, m1 remains fixed to the box. For k = 1, it returns with the same speed, v1. In the future, we will get devices with k = 0.95,… 0.98,… 0.99 and even more. This return coefficient will play an important role in the history of humankind.

"In figure 2.9, we show that we are restoring the velocity by injecting the quantity (1-k) * v1. We check speed restored, and we have

v1 = (1- k) * v1 + k * v1 = v1

"An engine supplies energy to the mass,

m1: E1f = v1^2 * m1/2

"On recovery, the engine receives energy.

E1r = (v1 * k)^2 * m1 * ½

"The energy we consume is energy supplied minus recovered energy. It's lossmaking energy. The higher the k value, with the less we consume, we compensate for lost energy (Ec). For that, I say this is multiple uses of energy.

Ec = E1f-E1r = (1-k^2) * v1^2 * m1/2

Ec = (1-k^2) * 100^2 * 40/2 = (1-k^2) * 200000 J

"With obtained data, we make a table with k variable for the following data.

m1 = 40kg, m2 = 2000kg (vehicle plus 4 passengers), v1 = 100m/s, T = 0.1s, F = 40 * 100/0.1 = 40000N

Force = p/T, impulse divided by period T

Power = E/T, energy divided by period T

Power table. F = 40000N

Nr	k^2	$(1-k^2)$	P = Ec/T [W]	F/P [N/W]
1	0.9	0.1	200,000	0.2
2	0.95	0.05	100,000	0.4
3	0.98	0.02	40,000	1
4	0.99	0.01	20,000	2
5	0.999	0.001	2000	20

Table 2

"After many calculations, I came to a summary table showing us great things," Ahve says. "We see that as k^2 approaches the value 1, the required power decreases. If we take k^2 = 0.999, we have a gain of 20 N/W. An athlete can develop about 400W. It can get 20N/W * 400W = 8000N (800kgf). We think we may put him like satellite around the earth. It's extraordinary! Keep in mind that we need a 0.999 return. With imagination, in the future, we will have a feeling of enthusiasm."

"I see that instead of using engines on vehicles," Eohn says, "we could horses or other traction animals. A horse with about 1000W can give 20000N."

"How can we achieve such large coefficients as 0.98, 0.99?" Reader asks.

"We have mechanical transmissions where power transmission is 0.98, 0.99," I say. "They do not have speed reflection, but we have it as a reference. In nature, we also have phenomena with higher coefficients. You see that I have presented some of the extraordinary phenomena that belong to a type-two civilization: propulsion generated by the vehicle without interaction with the environment and without evacuating mass. We can reuse the same energy, and that will benefit humankind greatly. But we must move mass centers very cautiously."

"About energy," Ahve says, "we have fossil fuels, which pollute and will be depleted. Hydroelectric, wind power, and solar panels are nonpolluting, but they can cause environmental damage.

Multiple uses of the same energy allow the extraction of energy from the surrounding space. It is nonpolluting, compact, and does not disturb the environment. Other advanced concepts will emerge in the future."

CHAPTER 3

STUDY DEVICES

"We're going to study some simple devices," I say, "and then deal with the next concept—how to compute and handle propulsion-related elements. These devices do not give any useful force. They are only for this exercise."

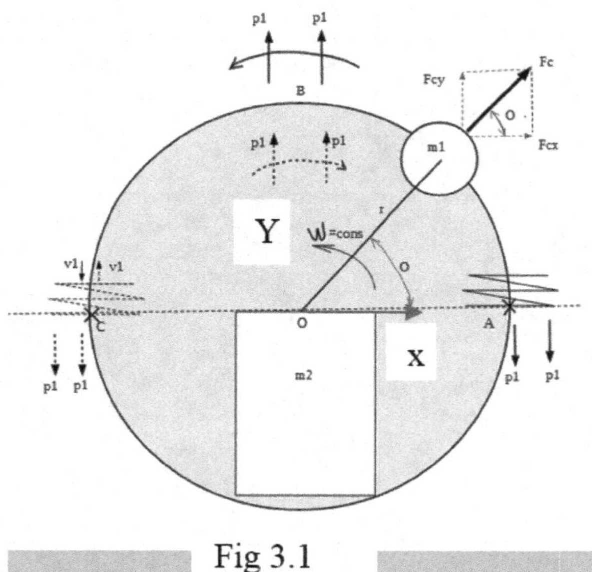

Fig 3.1

"I see we have some things to learn," Reader says, "by doing some calculations and experiments."

OSCILLATION AT A CONSTANT SPEED IN A SEMICIRCLE

"Let's consider a simple case," I say. "It's an exercise for future calculations. A mass rotates back and forth at a uniform speed on a 180gr arc. In figure 3.1. mass m1 moves on an arc of 180gr or pi radians. It departs from A with constant angular velocity; ω = constant. From C, it returns at the same speed. Mass m1 generates a centrifugal force: Fc.

Fc = ω2 * r * m1 [N]

"ω = is the rotational speed, angular in rad/s; r = is the radius from center O to m1; m1 is the mass (kg) in rotation.

"We design the centrifugal force Fc on the axes OX and OY.

Fcx = Fc * cos(Θ)   Θ = ω * t

Fcy = Fc * sin(Θ)   TAB = π/ω

"We are interested in the impulse given by Fc. It is the contribution of force to the movement of a mass. Two important factors are the magnitude of the force F and the time acting t, T.

p = F * t

"It can be decomposed, projected on coordinate axes.

$$pcx = \int_0^T Fc * \cos(\theta) * \mathrm{dt}$$

"This give us the impulse on the OX axis.

$$pcy = \int_0^T Fc * sin(\theta) * dt.$$

"This give us the impulse on the OY axis.

"We'll take a simple numerical example.

m1 = 10kg, ω = 1rad/s, T = π/ω = 3.14, is the time, the time from A to C, r = 1m

$$pcx = \int_0^\pi \omega^2 * r * m1 * cos(\omega * t) * dt = \int_0^\pi 1^2 * 1 * 10 * cos(1 * t) * dt = 0$$

"The resultant pulse pcx for t = π = 3.14 and the arc ABC, is zero.

$$pcy = \int_0^{\pi} \omega^2 * r * m1 * sin(\omega * t) * dt = \int_0^{\pi} 1^2 * 1 * 10 * sin(1 * t) * dt = 20\, Ns$$

"The resulting pcy pulse is 20Ns.

"The resort of C returns at a speed v1 mass m1. Impulses appear

p1 + p1 = 2 * p1 = 2 * 10 * 1 = 20Ns

"The sum of the impulses will be zero.

Pcx - 2 * p1 = 20 - 20 = 0

"The common mass center for m1 and m2 is immobile, but we have an oscillation of moving masses. It is an exercise and an introduction to the analysis of these mechanisms. You can make graphic representations that are interesting."

"I forgot the integrals you used to calculate the impulses," Reader says.

"Computers can do that, and books contain that information," Ahve says.

UNIFORM ROTATION AND RADIAL ACCELERATION

"We have one of the remarkable phenomena that occurs almost the same in nature," I say. "We study the simplest model. You exercise the description, drawing, and formulas to get into the phenomenon as much as possible. Put yourself in the middle of the experiment to perceive with your senses and develop your intuition. Get deep in your mind and senses the phenomena described through formulas and experiences. See what changes happen, and give them deep meaning. This is achieved through many exercises and practice.

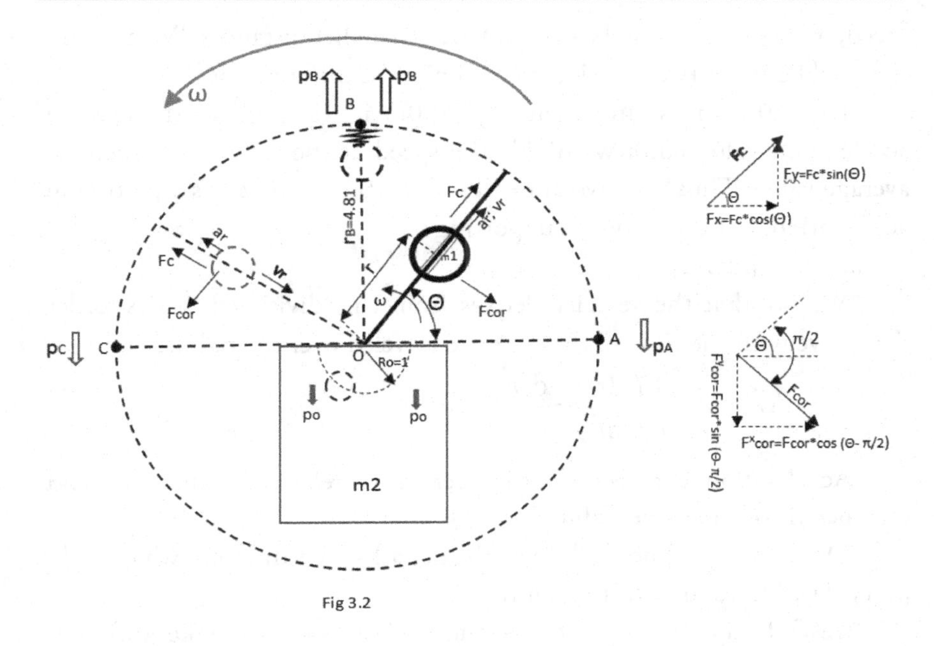

Fig 3.2

"In figure 3.2, we have an OM arm that rotates at a constant speed. The mass, m1, which moves radially, slides on it. The centrifugal force Fc pushes them from the center to the periphery. On the circle arc AB, mass m1 moves away from the center. In B, a resort returns at the same speed vr, the mass m1. On the arc BC, mass m1 moves to center O, and the centrifugal force decreases the vr to zero at C. Mass m1 moves in a circle with a constant radius Ro = 1 on the CDA arc. It generates a small centrifugal force. When it arrives at A, mass m1 is released and starts another cycle.

This is an average speed. If we want to find out the instantaneous speed, we take the smaller time 1s., 0.1s, 0.01s, and small distances. In other words, we measure the road between two positions r1 and r2. The difference is d, the differentials:

$vr = dr/dt = (r2 - r1)/(t2 - t1)$."

"A moving body passes through space in time," Ahve says. "We divide space by time and we have the speed. That tells us how many meters are traveled in one second. In our case, we are dealing with radial velocity, vr = r/t [m/s]. This is an average speed. If we want to find out instantaneous speed, we take short time 1s, 0.1s, 0.01s, and smaller distances. We measure the road between two positions rl and r2. The difference is d.

"I go 10m in 1s, and then I go 100m in 1s. To find the average acceleration value (am): we divide the speed at the time, so we get the average value. Thus we have am = (100-10)/10 = 9m/s^2; it's a pretty big acceleration, not everybody's supporting it.

ar = vr/t [m/s^2]

"We say that the velocity derives from this division is a first-order derivative with the time of the distance traveled over the radius.

$$ar = dvr \, / \, dt = \frac{\mathrm{d}}{\mathrm{d}t}\left(\frac{\mathrm{d}r}{\mathrm{d}t}\right) = \frac{\mathrm{d}^2 r}{\mathrm{d}t^2})$$

"Acceleration is a second-order derivative relative to the time and distance traveled on the radius."

"Do we need to know all this?" Reader asks. "It's different what we do every day. I have to reflect on all this."

"Yes," I say. "Computing is the easiest way to make different experiences. If there are no mistakes, we can know if we're on the right track. Then after the calculations, we are going to experiment and find out if the assumptions match the phenomenon studied. Calculations are exercises. Formulas express an idea. If you understand the concept, you can make different combinations. Practice. Let me tell you more. Let us consider centrifugal force. Put a 1kg mass, a bottle, at the end of a rope. Spin the bottle and feel the force generated by rotation—centrifugal force; it is a massicaly force.

"We perceive it as a punctual force along the cable," Ahve says. "Although each point is involved in its creation, we say it is a massic force. We do not feel the centrifugal force that balances gravity on a satellite of the earth as an example. In free fall, we do not feel the acceleration. It is useful to know when the forces are massic, surface, or punctual."

"Let me tell you about another fascinating, magical, mysterious force," I say, "the Coriolis force. It is perpendicular to radial movement. It results from a rotation movement with angular velocity ω and radial displacement with vr speed. The formula is Fcor = 2 * ω * vr * m. It's harder to see, but through it, we can explain phenomena from the atom to the motions in the cosmos. Seek to understand it as it will be of great help to us."

"How can we learn more about it?" Reader asks.

"Put a vacuum cleaner pipe in water or a pipe from a pen in a sink and turn it energetically to right, and to left. That takes a certain skill. Notice how water will come out the top as a result of centrifugal force. Turning the tube to one side and other, we feel the resistance that opposes it. Resistance is a force, the Coriolis force, and the momentum forctoward the center of rotation. With our hands on the pipe, we have the perception of force and momentum force. Try to distinguish between them. Do this experiment several times and you will notice that you come in time to more-advanced understanding. It's like a daily workout for athletes, musicians, and other professionals."

"Take a wire of a bicycle wheel or a thread," Eohn says. "Put on it a nut or a ring. By oscillation, you see how it goes up or down. They are simple devices that you should not miss out on; you can learn a lot from them. Use them and practice them and you will gain understanding of these concepts."

"But phenomena like these are mechanical. Why did we need them?" Reader asks.

"There are some details worth reviewing," Eohn says. "Who learns about the Coriolis force? We can live even if we ignore it, but it is still important."

"But can't we just read about them?" Reader asks.

"You will not understand in depth the phenomena described," Eohn says. "The formula is an idea. A calculation is the development of ideas, a road to a goal. If you do not know basic things, you will be lost and not understand much. You learn them by experimenting with them."

"We study the phenomena in figure 3.2," I say. "The sequence is as follows. We describe the event and see what the parameters are, we find the acceleration, the forces, and others, we write the mathematical relations of the parameters of the experiment, we process and develop the formulas to find the desired elements, and in the end, we draw conclusions from the final relationships."

"The idea seems simple," Reader says. "Give us the details of when you make calculations."

"The most important part is the description of the phenomenon," I say. "We attach to it the mathematical part, the relationship between the parameters, mathematical development, a demonstration of what leads us to draw conclusions. This process is a simplified simulation of the physical event we validate by experiment. We can easily mistake the calculations—we can miss a sign, we drop or forget a parameter, and so on. The most confusing thing is to make the wrong relationships that describe the phenomenon.

"In figure 3.2, a lever revolves with the velocity ω = constant. In rotation displacements, we have a set of equations or mathematical relationships. We have two visions: one for rotational movement and the other for linear movement. The two cases give the real movement. In rotational movement, the angle is θ [rad] and the angular velocity is ω [rad/s]. Multiplied by the time considered, t, gives Θ.

$\Theta = \omega * t$

"As units of measure, we have: Θ [radians] or Θ [rad]; at π = ~ 3.14 rad corresponds to an angle of 180 degrees. At π/2 ~ 1.57 rad, we have a 90 degree right angle. Computing machines are made to work with angular sizes in rad. Angular velocity ω is measured in rad/s.

"Besides the angular, circular speed, we have the linear vr or vr speed calculated in m/s. There is also accelerated radial movement due to the ar centrifugal force calculated as m/s^2.

"As you see, we speak of forces, accelerations. If we multiply the acceleration by the mass, we get the force. In calculations, we prefer to use ar because the formula is simpler. We have kinematics of movement.

$ar = \omega^2 * r\,[m/s\,\hat{}\,^2]$

"The formula is the starting point for our study. We place centrifugal acceleration in differential form.

$$ar = \frac{\mathrm{d}^2 r}{\mathrm{d}t^2}$$

$$\frac{\mathrm{d}^2 r}{\mathrm{d}t^2} = \omega^2 * r$$

"It is a differential equation that we solve. What do we want to find out?"

"We want to know the kinematics of the movement," Ahve says, "the expressions, values for r, vr, and ar and their connection with Θ and ω. We want to find out how the forces work. We make the sum; we were looking if we have resulted in a different from zero."

"We also make the algebraic amount of impulses," Eohn says. We want the solution, the function that satisfies the differential equation

$$\frac{d^2 r}{dt^2} = \omega^2 * r$$

"Clever mathematicians have found that the solution is of the form $r = e^{\alpha t}$, we want to find α = ? We derive the expression by r.

$$v_r = \frac{d}{dt} e^{\alpha t} = \alpha e^{\alpha t}$$

"As you see is the radial speed,

$$\frac{d^2 r}{dt^2} = a_r = \frac{d}{dt}\left(\alpha e^{\alpha t}\right) = \alpha^2 e^{\alpha t}$$

"We reached the radial, centrifugal acceleration.

"We process the above results.

$$\alpha^2 \cancel{e^{\alpha t}} = \omega^2 * \cancel{e^{\alpha t}}$$

"You see that we have made a simplification and thus we obtain this.

$$\alpha^2 = \omega^2$$

"We got the value for a.

$$\alpha = \sqrt{\omega^2} = \pm\omega$$

"The idea is that we have reached the value wanted for α. We can write the general equation of the sought equation this way.

$$r = C_1 e^{\omega t} + C_2 e^{-\omega t}$$

"This is the expression of the position along the radius of the circle of m1. I have found a formula that describes the radial position. We still want radial speed. It is derived with the time of r.

$$vr = \frac{dr}{dt}.$$

For acceleration, we have

$$a_r = \frac{dv_r}{dt}$$

$$v_r = \omega\left(C_1 e^{\omega t} - C_2 e^{-\omega t}\right)$$

$$a_r = \omega^2\left(C_1 e^{\omega t} + C_2 e^{-\omega t}\right)$$

"There are basic equations with which we will continue to operate."

"It seems we have many elements," Reader says. "How do we get into the essence of the phenomena?"

"I have the answer," Ahve says. "I understand that there are several elements and stages we go through to understand and make different combinations. We observe and describe the phenomenon. We make a physical pattern, a sketch. We correlate physical parameters—distances, speeds, accelerations, forces, pulses, and so on, and we link them in mathematical formulas for the above model.

"We then calculate the equations to reach our understanding and achieve future constructs. And then we experiment on a model, a prototype. From there, we can redo the whole process as often as necessary.

"To have a deeper understanding, we see in the formula as the physical phenomenon in its unfolding. When we calculate the equations, we see the changes in the studied events and understand the physical actions in the modifications. We try to feel them, to live in the experiment. We have to come to a profound understanding of the events studied to be able to handle the knowledge. We rise to the level of manipulating ideas, concepts for further development, and we see, feel, and understand the unfolding of the events."

"How could we use Einstein's E = mc^2 formula?" Reader asks. "We can have almost unlimited energy if we combine antimatter with matter and solve the energy problem."

"It's not like that," Eohn says. "You do not buy antimatter at a store. We do not have any immediate applications for it. Einstein's genius is that he saw the surrounding world with other eyes. He opened new ways of knowing."

"But we can also have new visions of the world," Reader says.

"Of course, these new concepts are in development. Let's see how far we can go. They are so performing that I'm hard to believe," Eohn says.

"Put together the basic relationships.

$$r = C_1 e^{\omega t} + C_2 e^{-\omega t}$$

$$v_r = \frac{dr}{dt} = \omega\left(C_1 e^{\omega t} - C_2 e^{-\omega t}\right)$$

$$ar = \frac{dv_r}{dt} = \omega^2 (C_1 e^{\omega t} + C_2 e^{-\omega t})$$

$$a\,cor = 2^* \omega^* vr$$

$$\theta = \omega * t$$

"We keep only a few of the relationships," I say, "r, the distance traveled along the radius; vr, the speed along the radius; ar, radial, centrifugal acceleration; acor, Coriolis acceleration; and Θ, the angle. The common elements are angular velocity ω and time t. Not all elements are presented such as moments of forces and impulses. Friction forces are not considered. We do not need them."

"If we multiply the mass with acceleration, we get the force," Ahve says. "The moment of a force is the force multiplied by the distance to the reference point. I understand that friction would complicate things, so we do not consider it. These simplifications do not affect the outcome."

"We want to see the contribution of the forces on the axes OX and OY," Eohn says. "Impulse is the contribution to the movement by forces over a determined time; it is achieved through integration. On axes, the projection is obtained by multiplying by the sine or cosine. We have to determine the coefficients C1 and C2. They are determined by the initial conditions. It can be seen that

$\omega * t = \Theta$

"So

$$\theta = 0,\ r(A) = Ro = 1m.\ vr(A) = 0; e^0 = e^{-0} = 1$$

$$r = C_1 e^0 + C_2 e^{-0} = 1;\ C1 + C2 = 1$$

"and

C1 = C2 = 1/2 = 0.5

$$v_r = \omega\left(C_1 e^0 - C_2 e^{-0}\right) = 0\,C1 - C2 = 0$$

"The equations for the above data are,

$$r = 0.5\left(e^{1*t} + e^{-1*t}\right)$$

"The value in B is

$$r = 0.5\left(e^{1.57} + e^{-1.57}\right) = 2.50918m$$

$$v_r = 0.5\left(e^{t} - e^{-t}\right)$$

The value in B is

$$v_r = 0.5\left(e^{1.57} - e^{-1.57}\right) = 2.3013m/s$$

$$a_r = 0.5\left(e^{t} + e^{-t}\right)$$

acor = 2 * 1 * vr

Θ = 1 * t

"We take values to simplify the calculations.

m1 = 1kg, ω = 1rad/s

tAB = π/(2 * ω) = 3.14/(2 * 1) = 1.57s.

"That is the time from A to B. The impulse on the OX axis, on the AB distance is this.

$$p^{x}cor = p^{x}cor = \int_{0}^{\pi/2} 2*1*0.5\left(e^{t} - e^{-t}\right)*\cos\left(1*t - \frac{\pi}{2}\right)*dt = 2.509$$

"The impulse on the OY axis, on the AB distance is this.

$$p^{y}cor = p^{y}cor = \int_{0}^{\pi/2} 2*1*0.5\left(e^{t} - e^{-t}\right)*\sin\left(1*t - \frac{\pi}{2}\right)*dt = -1.3013$$

"On the BC interval, we must have the equations of motion. This is done by determining the coefficients C1 and C2.

$$vr(B) = -2.3013m/s;\ r(B) = 2.50918m$$

C1 * 4.8104 + C2 * 0.20787 = 2.5091

1 * (C1 * 4.8104 - C2 * 0.20787) = -2.3013

"and

C1 = 0.02159, C2 = 11.57069

$$r = 0.02159*e^{1*t} + 11.57069*e^{-1*t};\ r(C) = 0.999 \cong 1$$

$$vr = 0.02159*e^{1*t} - 11.57069*e^{-1*t};\ vr(C) = -0.0004 \cong 0$$

"The values for the coefficients are correct. Mass returns to a position symmetrical with that at departure point A."

"Why do we change the equations?" Reader asks.

"Equations are the mathematical mirror of physical movement," Ahve says. "Each movement has its own equations. It is easy to track the movement through equations, graphics, and simulations. In fact, it is a simplified vision. We take into account the factors we believe are dominant. Experience tells us the truth. The measurements we compare with the calculations; if they are close, we say we are on the right path. Otherwise, we would start from the beginning."

"How close should the results be?" Reader asks.

"If the results are seventy percent, I did a good job," Ahve says. "Over seventy percent, the theory is very good."

"I see that mathematics is a kind of soul of phenomena," Reader says. "We have the idea, the project, and then the object, the matter. The image of things made by humans is put on computers: cars, buildings, airplanes, machinery, roads, bridges, and more. People have made stunning progress."

"We live in an imaginary world on our computers," Eohn says. "It is useful for various projects. Paper-based or computer-based projects are nothing but aids to planned work, not the soul of things. The soul is the bond between mortal creatures and eternity. I understand that in the future, we will be able to contact more-advanced entities."

"Let's continue with BC," I say. The impulse of Coriolis on the OX axis is this.

$$p^x cor = \int_0^{\pi/2} 2*1*(0.02159*e^{\frac{\pi}{2}+1*t} - 11.57069*e^{-\left(1*t+\frac{\pi}{2}\right)})*\cos\left(1*t-\frac{\pi}{2}+\frac{\pi}{2}\right)*dt = -2.50955$$

"The impulse of Coriolis on the OY axis is this.

$$p^y cor = \int_0^{\pi/2} 2*1*(0.02159*e^{\frac{\pi}{2}+1*t} - 11.57069*e^{-\left(1*t+\frac{\pi}{2}\right)})*\sin\left(1*t-\frac{\pi}{2}+\frac{\pi}{2}\right)*dt = -1.30103$$

"Hurray!" Reader says. "We have an impulse (B) = 2 * 2.3013 Ns greater than 2 * (- 1.3) Ns given by Coriolis forces. So we have propulsion."

"Keep your enthusiasm," I say. "It often happens. We are in a hurry, there are elements that escape us, and we count mistakenly. We stay for a while with the joy of apparent success. Then there are doubts that bring us back to redoing the calculations. We discover where we went wrong and go further. What else can we do? I did not show the mass movement on the CDA arc. Mass moves on the radius Ro = 1m with constant velocity ω = 1rad/s.

"On the OX axis, we have the impulse given by centrifugal force (cf).

$$p^x cf = p^x cf = \int_{2\pi}^{\pi} 1^2 * 1 * \cos(1 * t) * dt = 0$$

"On the OY axis, we have the impulse given by centrifugal force (cf).

$$p^y cf = \int_{2\pi}^{\pi} 1^2 * 1 * \sin(1 * \mathrm{t}) * dt = -2$$

"We make the sum of impulses on the axes.

$$p^y = 2 * (-1.3013) + 2 * (2.3013) - 2 = 0$$

$$px = 2.509 - 2.50955 = -0.00055 \cong 0$$

"There are small differences depending on the number of decimal places.

"In conclusion: the sum of impulses is zero; we have no movement. There are oscillations that are not useful to us. The mass center remains immobile.

"Remember and redo the calculations to deepen the phenomena; then once you understand, you change the data, do other exercises. We'll move on."

"Wait a minute," Ahve says. "This is a remarkable phenomenon. It would be useful for centrifugal pumps, power converters, water-based generators, moving air, and much more."

"We could extend this to the movements of atoms, satellites, planets, and beyond," Eohn says.

"I see you've caught something from the succession of ideas you've seen." I say. "You comprehend the development of some concepts. Keep your enthusiasm for what's next. When we obtain and operate with inertial traction, say, propulsion, applications are extraordinary; it's an attempt to get closer to a type-two civilization."

"We're all eyes and ears," Ahve says.

"Here is a brief anticipation of what will come," I say. "This experiment is a human adventure that goes far beyond what people have done, more than terrible wars, atomic explosions, catastrophes, and so on. By developing these, we will travel to other stars and guarantee water, food, and energy for all living creatures on earth."

"We can make heaven on earth," Reader says. "There will be no need for conflict."

"These extraordinary achievements can serve those who exploit savage people and the rest of the creatures until they are extinct," Eohn says.

"We should trust people," Ahve says. "They can eliminate deadly conflicts and extinction. We need high-level education and perfect ethics. People spend their energy on sports, art, space exploration, and more. I see a brake on human excesses—extraterrestrial civilizations can intervene vigorously. They can come to punish us or teach us. Let's reveal the secrets of nature and use them wisely."

ANGULAR ACCELERATION AND RADIAL ACCELERATION

"In figure 3.3," I say, "the OM lever moves from A by an accelerated angular motion e or Ɛ to B. From B it has a decelerated motion up to C.

"During this time, the mass m1 under the action of the radial forces has an accelerated motion up to B; here produces the impulse 2 * pB. From B, the radial speed changes its direction toward O. The speed decreases under the action of the radial accelerations; it is zero at C. From here, the cycle is repeated in the opposite direction.

"Remember! We have accelerated movements:

angular acceleration Ɛ

linear acceleration along the radius ar

Working with these two types of acceleration is essential. That is a subtlety that you should write down. Without these two accelerations, we do not have a positive result. Experiment with a pencil, a rod with a ring on it; rotate it manually and reflect on the phenomena."

Mathematical Model

"The angular acceleration e or Ɛ [rad/s^2] drives the movement as in the figure.

"The angular velocity: ω = Ɛ * t [rad/s] is the product of the t, movement time, and angular acceleration Ɛ.

The angle O or Θ is

$\Theta = \varepsilon * t^2/2$ [rad] Or $\Theta = \varepsilon * t^2/2 + \omega o * t + \Theta o$

The initial velocity is ωo and initial angle Θo. These three elements relate to rotation; we must use them carefully."

LINEAR MOTION

"Here we have some problems," I say. "We want to find the formulas for r distances, vr speeds, and ar accelerations. How can we?"

"We proceed as in the previous case," Ahve says. "We say radial acceleration is equal to centrifugal acceleration.

$$\frac{d^2r}{dt^2} = \omega^2 * r;\ replace\ \omega=\epsilon * t$$

"Solve the equation

$$\frac{d^2r}{dt^2} = (\varepsilon * t)^2 * r = \varepsilon^2 * t^2 * r.$$

"We try to solve it."

"It is an equation that has no solution, a primitive, a formula," I say. "Here, t^2, we see the variable tangled up in our calculations."

"We give it to mathematicians to solve," Ahve says.

"We solve it numerically or look for something else," Eohn says.

"You can solve it," I say. "I will take you through the steps. As you said, we were looking for something else. What are you looking for?"

"A formula," Ahve says.

"I'm looking for another idea, another phenomenon," Eohn says. "We drop the degree of the differential equation."

"Bravo!" I say. "We operate with phenomena, ideas we put into the formulas. Let's do an experiment. Let's cut a circle of cardboard and put a nail or toothpick in its center as an axis. We make it spin and put a drop of ink on the shaft. A spiral appears on the disc. On the spiral, we draw a tangent to the curve. We want the angle with the radius and the angle of the velocity to the tangent."

"We experimented. We obtained the angle of 45gr (π/4)," Eohn says.

"He gave me π/4," Ahve says.

"We obtained several similar spirals," Reader says. "I have angles between 42g and 47gr. I still have work to do. I think everyone needs to experiment."

"What is the ratio between speeds?" I ask.

"It is 1:1, the radial velocity; vr = dr/dt is equal to the tangential velocity, vt" Ahve says.

$vt = \omega * r = \mathcal{E} * t * r = dr/dt$

"If the relationship is not right, the whole construction collapses. It's a critical moment," Reader says.

"That's right," I say. "We have arguments to support it. The previous experience you have done confirms this. From figure 3.2, the relations show that for Θ, those who increase the tangential speed become equal to the radial one. I have conducted experiments that confirm the phenomenon."

"If the relationship is not in line with reality, we recommence the job." Ahve says. "We have a goal to find another suitable one. I think this one works."

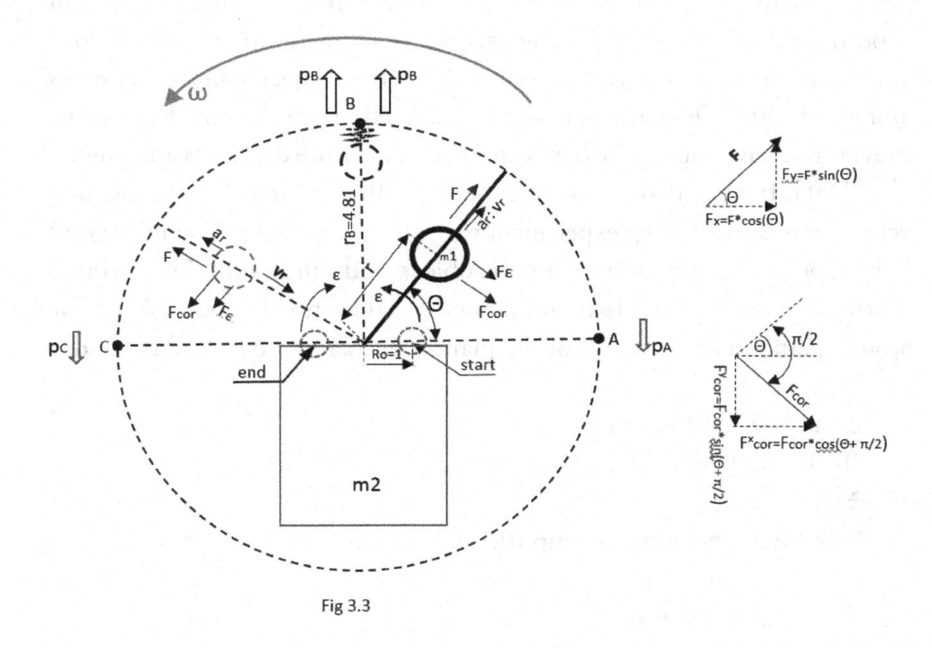

Fig 3.3

"I'm trying to introduce you to a more advanced understanding," I say. "Let's take the Coriolis acceleration

$acor = 2 * \omega * vr$.

We express that the radial velocity is equal to the tangential velocity: $vr = vt$ or $vr = \omega * r$, the formula is this.

$acor = 2 * \omega * vr = 2 * \omega * \omega * r = 2 * \omega^2 * r = 2 * acf.$

"The Coriolis acceleration (acor) is twice as high as centrifugal (acf) acceleration. Reflect on that. Perpendicular to the radius, Fcor is twice as big as Fcf over the radius."

"But the accelerations are unequal, different," Reader says. "How do we find the relationships between them?"

"Practice," I say. "You have many problems with the case in figure 3.2. You will see that $vr \cong vt$ for angles over $\pi/2$. It is a natural tendency for speeds to equalize. If they are not equal, the relationship between them is expressed by exponential functions; this happens at the beginning of the movement."

"Can we experimentally check this?" Reader asks.

"Of course," I say. "I can describe a simple, cheap experiment that everyone can do with their equipment. Take a movie camera capable of shooting 30, 60, 120 images per second. Stay on the first floor. Below, your assistant turns a lever, a pipe, a stick, with an object, a mass that slides along its length. On it attach a wheel graduated in centimeters. Review the movie. You have equal time between images; measure distances and angles.

"Data from calculations compare to linear, angular, accelerated velocities measured. The experiment reproduces figure 3.3. You need good light, a perpendicular position on the object, and a little skill. I urge you to do this and realize you are reproducing partial events that occur from the atomic plane to the motions of the planets in solar systems and beyond.

"Return to

$2 * \omega * vr = 2 * \omega * \omega * r$

"If this is the formula,

$\cancel{2} * \omega * vr = \cancel{2} * \omega * \omega * r$

"We made the marked simplifications. We get

$vr = \omega * r \; ; \; \omega = \varepsilon * t$

"Keep in mind this.

$$vr = \frac{dr}{dt} \text{ and } \varepsilon * t = \omega$$

$$\frac{dr}{dt} = \varepsilon * t * r$$

"This follows a separation of the variables

$$\frac{dr}{r} = \varepsilon * t * dt$$

"It is an easy-to-integrate equation; it has the solution.

$$\ln(r) = \varepsilon \cdot \frac{t^2}{2} + C_0$$

"where the value for r is

$$r = C * e^{\frac{\varepsilon t^2}{2}}$$

"where e = 2,7182 … A constant.
"From the initial conditions: for t = 0, r = Ro,

$$Ro = C * e\text{^}0$$

"where C = Ro. The final equation is

$$r = Ro * e^{\frac{\varepsilon t^2}{2}}$$

"This is the solution sought. We note that angle
θ = (ε·t^2)/2; the radial displacement r has the form
r = e^{θ}, where θ is the angle traveled.
"We have detailed the above calculation for you have to see details.

We followed a path similar to the previous case. We want to examine the impulses on the OX and OY axes.

"We have the following set of equations we operate with. For angular displacements

1) Ɛ = constant = ± (1 … n)

"n is real positive number for angular accelerations; negative for angular decelerations; units of measure (rad/s^2).

2) ω = Ɛ * t

"This is the angular velocity accumulated after time t; units of measure (rad/s)

3) θ = (ε * t^2)/2, θ is the angle traveled after time t; units of measure (rad)

"For

$$\varepsilon = 1, \theta = \pi / 2; t(\pi / 2) = \sqrt{\frac{2\theta}{\varepsilon}} = \sqrt{\frac{2\pi}{2 \cdot 1}} = 1.77245\ldots s$$

"This is the time from A to B. This is for rotation motion.

"For linear displacements; along the radius,

4) $r = Ro * e^{\frac{\varepsilon t^2}{2}}$

"This the radius of m1 for the angle $\theta = \frac{\varepsilon \cdot t^2}{2}$; *or* r = e^{θ}

"We highlight the angle of rotation.

For

$$\varepsilon = 1, \theta = \pi / 2, \ Ro = 1; r(B) = 1 * e^{\frac{\pi}{2}} = 4.8104 \ldots \ m$$

5) $Vr = Ro * \varepsilon * t * e^{\frac{\varepsilon t^2}{2}}$.

"This is the radial velocity $vr = \frac{dr}{r}$; *or* $Vr = Ro* \omega * e^{\theta}$. Speed in B, Vr(B) = 1 * 1 * 1.77 * $e^{(\pi/2)}$ = 8.5263 … m/s

6) $ar = Ro * \varepsilon^2 * t^2 * e^{\frac{\varepsilon t^2}{2}} = Ro * \varepsilon * e^{\frac{\varepsilon t^2}{2}}$.

"This is the radial acceleration.

7) $a\varepsilon = Ro * \varepsilon * e^{\frac{\varepsilon t^2}{2}} + Ro * \varepsilon * e^{\theta}$.,

"This is the tangential acceleration, is parallel to the Coriolis acceleration.

"These seven equations are the basis on which we will build our construction. There is still a force that will appear later."

"Do we have to remember these? They're complicated," Reader says.

"No formula must be memorized," Ahve says. "You have to know the logic that leads to it, the calculations, and understanding. The most important is to feel the phenomenon intimately."

"Our work opens many discussions in all areas of life," I say. "Let's go back to our calculations. We calculate the impulses on the axes OX and OY.

"The Coriolis impulse is

m1 = 1kg, Ɛ = 1 rad/s^2, Ro = 1m, t (B) = 1.7724 s

$$p^x cor = \int_0^{1.772} 1 * 2 * \varepsilon * t^1 * Ro * \varepsilon * t * e^{\frac{\varepsilon t^2}{2}} * cos(\frac{\varepsilon \cdot t^2}{2} - \pi / 2) * dt = 8.6014 Ns;$$

for the OX axis.

$$p^y cor = \int_0^{1.772} 1 * 2 * \varepsilon * t^1 * Ro * \varepsilon * t * e^{\frac{\varepsilon t^2}{2}} * sin(\frac{\varepsilon \cdot t^2}{2} - \pi / 2) * dt = -4.3463 \ Ns$$

for the OY axis.

"The impulse for tangential inertial forces are

$$p^x\varepsilon = \int_0^{1.772} 1 * Ro * \varepsilon * e^{\frac{\varepsilon t^2}{2}} * cos(\frac{\varepsilon \cdot t^2}{2} - \pi / 2) * dt = 2.05136\ Ns;$$

for the OX axis

$$p^y\varepsilon = \int_0^{1.772} 1 * Ro * \varepsilon * e^{\frac{\varepsilon t^2}{2}} * sin(\frac{\varepsilon \cdot t^2}{2} - \pi / 2) * dt = -2.1283\ Ns;$$

for the OY axis

"It is enough to make the py pulse balance.

"On the ABC arc, the px pulses on OX are canceled due to symmetry. We will see it further.

py = pycor + pyƐ + m1 * vB = -4.3463 -2.1283 + 1 * 8.5263 = **2.0517** Ns

"With this partial calculation on the arc AB, we have a sensational result! It shows movement on the OY axis. The center of mass moves. That is not possible in the current technology taking into account the vehicle and masses of expelled matter."

"I'm not convinced that from B to C we have symmetry for impulses. You go too fast," Reader says.

"I'll give you the BC calculation," I say. "Try to rebuild formulas yourselves. It's like having partial information from another civilization. You want to understand, develop, and use them. Stop reading and doing the calculations. If you know the phenomena, the formulas, the demonstrations, try to do the rest of your calculations yourself. Test yourself this way. For verification, I tell you that you get the above results due to symmetry.

"For circle arc BC, we have to pay attention to the parameters of the movement.

$$\theta + \pi / 2 = \omega o * t - \frac{\varepsilon \cdot t^2}{2} = \pi / 2 + 1.772 * t - \frac{\varepsilon \cdot t^2}{2}$$

"We see that for

t = 0, θ (B) = π/2

$r_B = 4.81\ m,\ t(BC) = 1.772 s, \omega o = 1.772 rad / s$

ω = ωo –t * Ɛ = 1.772-t * Ɛ.

"The velocity ω decreases in the shift from B to C.

$r = 4.81 * e^{-(1.772*t - \frac{\varepsilon t^2}{2})}$; is the value of the position (r) of m1 from the center of rotation O.

"The problem can be solved quickly with the help of the relationship

$$vr = vt = \omega * r = (\omega o - t * \varepsilon) * 4.81 * e^{-(1.772*t - \frac{\varepsilon t^2}{2})} = 4.81 * (1.77 - t * \varepsilon) * e^{-(1.772*t - \frac{\varepsilon t^2}{2})}$$

"A vision starts from the equality of the radial speed with the tangential speed. We note that the ω rotation speed is trigonometric in the way we designed this device.

$$vr = -dr / dt = 4.81 * (1.772\ -) * e^{-(1.772*t - \frac{\varepsilon t^2}{2})}$$

"It is the same but expressed differently by derivation. Speed is positive and position on radius r decreases. But I prefer the way to show you the forces of action and reaction.

"θ can have two values

$$\theta = \pi / 2 \pm \pi / 2 + 1.772 * t - \frac{\varepsilon \cdot t^2}{2}$$

Fcor = m1 * 2 * ω * vr

Let us remember the expression of the Coriolis force.

$$p^x cor(0) = \int_0^{1.772} 1*2*(1.772 - \varepsilon * t)*4.81*(1.772-)*$$

$$e^{-(1.772*t - \frac{\varepsilon t^2}{2})} * cos(1.772 * t - \frac{\varepsilon \cdot t^2}{2}) * dt = 8.59895\ Ns;$$

values for the OX axis, for $\theta = 1.772 * t - \frac{\varepsilon \cdot t^2}{2}$; it is the impulse on the OX of the force of action.

$$p^x cor(\pi) = \int_0^{1.772} 1*2*(1.772 - \varepsilon * t)*4.81*(1.772\varepsilon t-)*$$

$$e^{-(1.772*t - \frac{\varepsilon t^2}{2})} * cos(\pi + 1.772 * t - \frac{\varepsilon \cdot t^2}{2}) * dt = -8.59895\ Ns;$$

values for the OX axis; for $\theta = \pi / 2 + \pi / 2 + 1.772 * t - \frac{\varepsilon \cdot t^2}{2}$

"It is the impulse on the OX of the reaction force. It is the factor that drives our system. It is seen that the impulses are equal and opposite. See figure 2.1 where F = -F. We have equality of action and reaction in all of the examples below.

$$p^y cor(0) = \int_0^{1.772} 1*2*(1.772 - \varepsilon * t) * 4.81 * (1.772 - \varepsilon * t * e^{-(1.772*t--}$$

$$* sin(1.772 * t - \frac{\varepsilon \cdot t^2}{2}) * dt = 4.3425\ Ns;$$

values for the OY axis; for $\theta = 1.772 * t - \frac{\varepsilon \cdot t^2}{2}$; it is the impulse on the OY of the force of action.

$$p^y cor(\pi) = \int_0^{1.772} 1*2*(1.772 - \varepsilon * t)*4.81 * (1.772 - \varepsilon * t)$$

$$* e^{-(1.772*t - \frac{\varepsilon t^2)}{2}} * sin(\pi + 1.772 * t - \frac{\varepsilon \cdot t^2}{2}) * dt = -4.3425\ Ns;$$

values for the OY axis; for $\theta = \pi + 1.772 * t - \frac{\varepsilon \cdot t^2}{2}$; it is the impulse on the OY of the reaction force.

"For tangential forces,

$$p^x \varepsilon(0) = \int_0^{1.772} 1*4.81*(-\varepsilon)*e^{-(1.772*t - \frac{\varepsilon t^2)}{2}} * cos(1.772 * t - \frac{\varepsilon \cdot t^2}{2} * dt = -2.0526 Ns;$$

values for the OX axis; for $\theta = 1.772 * t - \frac{\varepsilon \cdot t^2}{2}$: it is the impulse on the OX of the reaction force.

$$p^x \varepsilon(\pi) = \int_0^{1.772} 1*4.81*(-\varepsilon)*e^{-(1.772*t - \frac{\varepsilon t^2)}{2}} * cos(\pi + 1.772 * t - \frac{\varepsilon \cdot t^2}{2}) * dt = 2.0526\ Ns;$$

valori pentru axa OX; for $\theta = \pi / 2 + \pi / 2 + 1.772 * t - \frac{\varepsilon \cdot t^2}{2}$; it is the impulse on the OX of the force of action.

$$p^y \varepsilon(0) = \int_0^{1.772} 1*4.81*\varepsilon*e^{-(1.772*t - \frac{\varepsilon t^2)}{2}} * sin(1.772 * t - \frac{\varepsilon \cdot t^2}{2} * dt = 2.12825\ Ns;$$

values for the OY axis; for $\theta = 1.772 * t - \frac{\varepsilon \cdot t^2}{2}$; it is the impulse on the OY of the force of action.

$$p^y \varepsilon(\pi) = \int_0^{1.772} 1*4.81*\varepsilon*e^{-(1.772*t - \frac{\varepsilon t^2)}{2}} * sin(\pi + 1.772 * t - \frac{\varepsilon \cdot t^2}{2}) * dt = -2.12825\ Ns;$$

values for the OY axis; for $\theta = \pi / 2 + \pi / 2 + 1.772 * t - \frac{\varepsilon \cdot t^2}{2}$ it is the impulse on the OY of the reaction force."

"If all the impulses given by the action and reaction are equal, what do we do with propulsion?" Reader asks. "I do not see that it can!"

"Did you redo the calculations above?" I ask. "Did you first understand the ideas and then the demonstrations?"

"I read them. I tried to do the calculations, but I did not insist," Reader says.

"I'll explain," Ahve says. "Do you eat bread?"

"Yes, I eat bread made from wheat," Reader says.

"So the wheat puts it in your mouth and chews it?" Ahve asks.

"You're laughing at me," Reader says. "Wheat goes through many processes to become bread."

"So different machines, tools, and others are made of fused, cleaned ores, transformed into slabs, wires, blocks that by other changes become machines, tools, etc. They are finished products," Ahve says. "We also have processes of energy, of the power that enters our devices; it is transformed, it is processed. We have forces, impulses, and we design impulses on axes. We use sines, cosines, and we have energy recovery. The result of this process is an impulse difference in OY in our case. This impulses difference in one second gives traction forces. Traction force is the product of this process just as bread is the final product of wheat."

"What is the difference between energy and power? Besides, I have not heard of nonmaterial products," Reader says.

"E = m * v^2/2 is the mechanical energy," Ahve says. "It is like the gasoline tank. The power is how much energy you consume in a second, P = dE/dt. A water drop can turn an electrical generator. That mechanical energy comes to you through wires in the form of electricity. It is a transformation you do not see, but it can be used. It has no visible physical form. Note that the force as the final product is obtained from energy processing."

"We make the impulses balance," I say.

$p^x = 8.6014 + 2.05136 - 8.59895 - 2.0526 = 0.000121 \sim 0$

"This is zero. It has no movement, but we see there is oscillation on OX.

$p^y = -4.3463 - 2.1283 + 2 * 8.5263 - 4.3425 - 2.12825 = 4.10725$ Ns

"In this, movement appears. ***I have demonstrated the existence of the impulse of traction—propulsion! We have movement! The mass center***

moves! In addition, is an oscillation on OY. Movement is intermittent. Here are some useful calculations for inertial traction. On AB, we have to inject an EAB energy that we calculate from the sum of radial and peripheral energies.

EAB = m1 * (vr^2/2 + vƐ^2/2) = 1 * (1.77^2/2 + 1.77^2/2) = 3.14 J

"We want to find the traction force.

$F_{oy} = p^y/(2 * T) = 4.10725/2 * 1.77 = 1.158$ N

"This is the force we get on OY. The average power for motion on AB is

PAB = EAB/(T) = 3.14 J/(1.77s) = 1.77 W

"We want to find out what power is required for the traction force; we consider a 0.9 (ninety percent) yield; that means we injected into the system at each cycle 0.1 (ten percent) to compensate for losses.

Foy/(PAB * 0.1) = 1.158 N/(0.1 * 1.77W) = 6.54 N/W.

"These results do not have big values. What do we do?" Reader asks.

"I find them good," Ahve says. "These are experiments to begin with. We can play with them and see what happens. They are slightly visible and a little dangerous. Different values can be changed by adjusting the motion parameters."

"I'll give you another stronger case," I say.

Ɛ = ±100rad/s^2

ω = Ɛ * t = 100 * t

$$t_A = 0; t_B = \pm\sqrt{\frac{2\pi}{4 \cdot 100}} = \pm 0.12533..s$$

$\theta_A = \pi/4$, $\theta_B = \pi/2$, $\theta_C = 3 * \pi/4$, $\theta = Ɛ * t\text{^}2/2 + \pi/4$;

$$r = Ro * e^{\frac{\varepsilon t^2}{2}} = 1 * e^{\frac{100t^2}{2}}; rB = 2.184..m$$

$$Vr = Ro * \varepsilon * t * e^{\frac{\varepsilon t^2}{2}} = 1 * 100 * t * e^{\frac{100t^2}{2}}; \ for\ t = 0.125s; Vr_B = 27.25\ m/s$$

"The impulse in B is

$pr_B = m1 * 27.25 = 100 * 27.25 = 2725$ Ns

$$p^x{}_{cor} = \int_0^{1.25} 100 * 2 * 100 * t * 1 * 100 * t * e^{\frac{100t^2}{2}}$$

$$* cos(100 * t\text{^}2/2 - \pi/2 + \pi/4) * dt = 2004.55 Ns$$

$$p^{y}{}_{cor} = \int_{0}^{1.25} 100*2*100*t*1*100*t*\mathrm{e}^{\frac{100t^2}{2}}$$

$$*sin(100*t\hat{}2/2-\pi/2+\pi/4)*dt = -567.28Ns$$

"For tangential forces, collinear forces with Coriolis we have,

$$p^{x}\varepsilon = \int_{0}^{1.25} 100*100*\mathrm{e}^{\frac{100t^2}{2}} *cos(100*t\hat{}2/2-\pi/2+\pi/4)*dt = 1444.31Ns$$

$$p^{y}\varepsilon = \int_{0}^{1.25} 100*2*100*t*1*100*t*\mathrm{e}^{\frac{100t^2}{2}}$$

$$*sin(100*t\hat{}2/2-\pi/2+\pi/4)*dt = -739.95Ns$$

"Taking into account the previous experiment, we can see that the impulses on the OX axis are canceled due to the symmetry: px = 0.

"On the OY axis, we have,

p^y = 2725 - 567.28 - 739.95 = 1417.77 Ns; for t_{AB} = 0.125s

"Notice that the impulse on OY has the weight of fifty-two percent (1417.77/2725 * 100 = 52.028%), is higher than the previous case that is twenty-four percent. (4.10725/2 * 8.5263 * 100 = 24.08%).

"The force on the OY axis is this.

Fy = p^y/tAB = 1417.77/0.125 = 11342.16N

"The energy injected on AB is this.

E_{AB} = m1 * 2 * vB^2/2 = 100 * 2 * 27.25^2/2 = 74256 J

"Considering a ninety percent recovery, the filling power is 0.1 (ten percent).

"For a ninety-eight percent recovery, the filling power is 0.02 (two percent).

P^{98} = 0.02 * E/t = 0.02 * 74256/0.125 = 11881W

"For the force/power ratio, we have the following cases.

F_y/P^{90} = 11342.16N/59405W = 0.2N/W

"For a ninety percent recovery, we consume 1W and get a 0.2N force.

F_y/P^{98} = 11342.16N/11881W = 0.95N/W

"For a ninety-eight percent recovery, we consume 1W and obtain a force of 0.9N.

"For 11330N, we have applications for land, water, air, and cosmos.

"From here, you can imagine different developments. I'll give you an example. A 600kg vehicle can be used in accelerated displacements with a = 18.9m/s^2. It can be raised vertically with a = (18.9-10) ~ 9m/s^2; we extract the gravitational acceleration. On acceleration times of 1000s + 1000s of deceleration, it moves through space:

2 * S = 2 * a * t ^{2}/2 forward; the distance is double.

S = 2 * 9 * 1000^2/2 = 9000000m = 9000km for a time of 33.3min (2000s), has an ascending speed of up to 9000m/s and then decreasing. An airplane needs nine to ten hours to fly that distance, twenty times longer."

"There are other issues," Eohn says. "What is the mass of this engine? We have shareholders, there is a casing, air reserves, and more."

"We do not design a device," I say. "We are at the level where we focus on developing the principle of operation. These are future stages. We still have to work."

"I still have my doubts," Reader says. "How do these lessons that I do not really understand let us draw conclusions that can turn the world upside down?"

"You are like Thomas the unbeliever," Ahve says. "Well-run calculations are like an experiment. They tell you if the phenomenon is true or not. I see it is fair. We have a good result."

"We can get more from this phenomenon," I say. "Do you want to make discoveries to find new laws?"

"Yes we do," Eohn says. "We'll become rich, and we want to participate, to see how these ideas are born, to feel the creative emotion."

"Don't be deluded," I say. "Most visionaries and inventors have died poor and sick. Others made the profit."

DISCOVERING NEW RADIAL ACCELERATIONS

"Let's see how we get these results. We take the basic equations and analyze them.

$$r = Ro * \mathrm{e}^{\frac{\varepsilon t^2}{2}}$$

"This is the distance along ray of m1.

$$Vr = Ro * \varepsilon * t * \mathrm{e}^{\frac{\varepsilon t^2}{2}} .\wedge$$

"is the radial velocity $vr = \frac{\mathrm{d}r}{r}$; *sau* $Vr = Ro* \omega*e^{\theta}$ ·

"Note that: ε * t is the angular velocity that multiplies the radius r.

"The advance in knowledge is the next equation, where we will make some changes to understand them easily.

$$ar = Ro * \varepsilon^2 * t^2 * e^{\frac{\varepsilon t^2}{2}} + Ro * \varepsilon * e^{\frac{\varepsilon t^2}{2}} \cdot$$

"is the radial acceleration.

$$ar = Ro * \omega^2 * e^{\theta} + Ro * \varepsilon * e^{\theta}$$

"is the same in another form. Say what the two terms are. Who sees it?"

"The first term is centrifugal acceleration

$$Fc = Ro * \omega^2 * e^{\theta}; Fc = \omega^2 * r$$

Ahve says. "We can spin objects to feel it. This force pushes the mass m1 along the radius. At the end of the movement, in B, we use the accumulated impulse. That is all."

"I'm trying my luck," Eohn says. "It seems to me to be tangential acceleration is radius multiplied angular acceleration."

"That's what I thought at first," I say, "but it's not like that. Discovery: F_{ε} is acceleration along the collinear radius with the centrifuge. It adds to it. The expression is

$$F_{\varepsilon} = \varepsilon * Ro * e^{\theta} = \varepsilon * r; where\ r = Ro * e^{\theta}$$

$$Ro * e^{\theta}$$

"It explains that we can get a bigger impulse on the OY axis. This force manifests itself from the atom to the cosmos. The law is just as old as the universe."

"I have a possible explanation for the cosmos," Ahve says. "I believe that the faster expansion of the universe is explained by the presence of force F_{ε}. It is Fc+FƐ=F. It accelerates faster than Fc only. For large masses, we have a radial velocity and get decelerate rotation (Ɛ) due to the Coriolis forces. So we have the conditions for F_{ε}; the masses have a double accelerated motion."

"Do you mean that the black-matter hypothesis is no longer necessary?" Eohn asks.

"I cannot say there is no black matter," Ahve says. "I do not have that knowledge. This discovery of F_{ε}, the radial force is given by angular acceleration is attractive, beautiful, and I think it can have many applications. A use for inertial traction is most important. It explains getting the extra boost on OY. It is great progress."

"Now I understand that there is something concrete, palpable that can give the pulling force," Reader says. "I'm not used to reading the formulas. I still understand that it is a very important achievement for the well-being of people and creatures. Now I'm quieter. Bravo! We can also cry 'Eureka!' as did Archimedes, who ran naked from his house after he made his discovery. You deserve a prize."

"The prize is like the lottery," I say. "Many give money and maybe one wins. If you'd like, we can continue. We have other discoveries. Miracles at first glance can be taken as miracles. What do you say?"

"Of course we agree," Eohn says. "The importance of the F_{ε} discovery, maybe we didn't understand it well. It takes time for thought."

DISCOVERING AN EXPLANATION OF INERTIA

"Notice that we have a system where when we introduce energy, we get a complex movement—it results in a force," I say. "We do the opposite. We apply a force (acceleration) to a mass that induces at the atomic level opposing movements generating a counterforce—inertia. So the hypothesis is that inertia is generated by movement of atomic elements of the body mass. The long-term idea will give you further developments."

"That's quite a statement," Eohn says. "To be credible, calculations and experiments should be made."

"We made calculations on a simple model, as in figure 3.1 modified," I say. "The result is positive. There is the experiment. This is not in my immediate attention. I gave it for the possible developments of the readers."

"This discovery can fill the law of Newton's inertia," Ahve says. "We say that the mass-induced movements generate the forces of inertia. It is an excellent gain and understanding. With this, the knowledge of classical mechanics progresses."

"We are making it complete in addition to the knowledge of classical mechanics," I say. "I hope to push it, to give deeper and more-accurate explanations of nature. As you see, we have put forth developments and ideas that other science does not have. Inertia is not a matter of immediate attention. So we move on."

CHAPTER 4

POWER GENERATORS

"So here we are creating a big bang," Reader says. "With energy and matter, we create a new universe. Ha! The fantasies dreamed of millenarians—the philosopher's stone, the transformation of lead into gold, and the elixir of life. I'm curious what stories you have."

"This man does not trust the power of reason, inspiration, and calculation," Ahve says. "He sees no benefit for the creatures on earth. He should be punished. And shut up!"

"We all have such weaknesses," I say. "We doubt, hesitate, do not go further because we are comfortable."

"If we start down a road, we should go to the end," Ahve says.

"That's how we get all misfortune on earth!" Eohn says.

"Let's leave the passionate dispute and continue," I say. "Tell me what we want. We were looking for this objective."

"We want energy, a gas tank, a bunch of wood," Reader says.

"We want force, movement," Ahve says.

"We want mechanical power to run electric generators to give the same power," Eohn says.

"Eohn is on the right path," I say. "So far, we have imagined the way to go. Let's get it. What is the formula we apply?"

"Power is force multiplied by its speed, $P = F * v$," Ahve says.

"How do you imagine this?" I ask. "What would be the construction?"

"A machine is powered by force F, has speed v, and a generator," Eohn says.

"More compactly, it is a force that rotates a generator with the peripheral speed v," Ahve says.

"A solution is presented in figure 4.1," I say. "Let's get to more details. We take the data from figure 3.3. We specify the parameters, and we consider two cases for comparison

F = 11342.16N

P90 = 0.1 * E/t = 59405W "and

P98 = 0.02 * E/t = 11881W

v = 4m/s peripheral speed. r = 4m peripheral radius

The power generated is Pg = F * v = 11342.16 * 4 = + 45368.64W

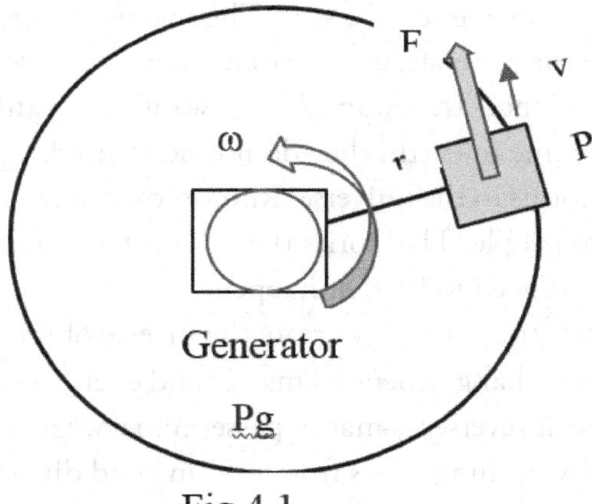

Fig 4.1

"The case in which we consume energy. The power to maintain the operation is negative.

Pg - P90 = 45368.64 - 59405 = -14036.32W

"This is a power we give to the system. No energy is produced. It consumes energy.

Pg - P98 = 45368.64 - 11881 = + 33487.64W

"This is generated energy we can use. In our case, we get three times higher energy. Note that we can make it consume or generate energy. How you see energy generation depends greatly on energy recovery; we can see speed return from B plus the rotation recovery."

"This thing leaves me with an open mouth," Reader says. "As you say, it looks like it works. The general opinion is that it cannot. I have to consult with others. What is the energy consumption?"

"Energy is consumed without it transforming into heat as when you brake a car," I say. "It's simply consumed. It also has a cosmic scale image."

"If we have a power generator, we can do what we want," Reader says. "According to Einstein, mass transforms into energy and vice versa "$E=mc^2$ "a mass transforms it into ene, so we can generate mass as much as we want. Maybe that's what happened in the big bang. An artificial device or a natural phenomenon may have initiated the process."

"Certain other civilizations know about accelerated inertial displacement, so we name it," I say. "They know the generation of energy and implicitly the generation of matter. They can have what metals they want, what substances they want. If they want, they can fabricate a planet, a sun. If they come to earth, they do not need anything. I think they are spreading life forms in the universe. Rumors of an alien invasion are meant to put fear into people. The stories that UFOs have crashed I say are false; they are too advanced to let that happen.

"Energy consumption may create the inverse of the big bang. It closes the cycle. The big bang generated matter and energy billions of years ago, but after a time, it reverses—matter passes into energy, it is consumed, and it disappears. Everything starts from nothing and dissolves into nothing."

"Are you sure?" Reader asks.

"These are working hypotheses," I say. "They help us advance. As we progress, we modify them according to reality. Imagine ideas or other concepts. Let's go further."

CHAPTER 5

DEPOLLUTION

"Artificial or natural waste obstructs the normal course of life," I tell the others. "We will modify the; we will turn them into useful compositions. In short, I present a hypothesis based on the concept of unlimited available energy.

"Thermal dissociation may be a radical form of waste passing into constituent chemical elements. They are returned in nature or are raw materials for another cycle of use. We use reactors with temperatures above 3000C for the dissociation of substances. The most difficult ones are CO, which needs more than 3670°C. CO_2 needs temperatures above 2000°C, and methane CH_4 above 2000°C. For help, we can use plasma, laser, microwave, and catalysts.

"In addition, we need reactors that are made of high-fusion metal. W, tungsten, melting point: 3695K (3422°C). Mo, molybdenum, melting point: 2896°K (2623°C). Ir, iridium, melting point: 2719°K (2446°C). Presumably, we develop metallurgy in the cosmos, where the lack of gravity makes it easier to manufacture metals.

"Other low-energy depurations are used on a case-by-case basis—sedimentation, fermentation, filtration, and so on.

"We are in a late stage, and we are not dealing with it. I open a discussion for events to come, and I need the help of readers.

CHAPTER 6

ETHICS: KEEPING ALL CREATURES ON EARTH AND IN THE COSMOS ALIVE

"We need more details, concepts, and ideas," I say. "What we are talking about here is just a starting point. I think all people can see the unfolding of events in their own ways. We call upon them to support us to protect their futures and those of their descendants."

"For life on earth or in the cosmos, we need to provide air, water, food, and optimal living environments," Ahve says. "If antelopes, lions, and elephants do not have water, we should bring it in the form of large blocks of ice."

"Ethics is at the highest level," Eohn says. "The means to achieve the goal, we will need all humanity's creations: strength, technology, money, finance, culture, and political organizations. They are the tools that can help achieve it."

CHAPTER 7

CRITICISM OF THE CURRENT SYSTEM

"We want to sketch the future through the new knowledge shown above," I say. "Everyone who reads it should talk with their friends and family. Imagine a balancing scale. On one side, we put the elements of the present. On the other side, we put the aspects of a type-two civilization. We play with them; we move items from one balance to the other. Balance is allegorical; it's in our minds. We judge, choose, develop, and act so life here and in the cosmos can flourish harmoniously. The unit of measure I propose is life. We can also choose other units of measurement—death, destruction, dictatorship, slavery, capitalism … What do you prefer?"

"I say we keep the good and useful and put the rest in museums," Ahve says, "so our descendants can see them. We should preserve for young people more cities, agricultural lands, and other past achievements. People should see and understand the past so they can compare it to their present."

"We are initiating criticism of today's society," I say. "This criticism should be as harsh as possible. The goal is to develop what is useful. We want to give up, eliminate, or mitigate the destructive ones but keep them in mind. Obviously, there is much to say. What you read may inspire you, bring new ideas, other visions.

"We compare the current situation with the likely future we imagine but do not know. We have a futuristic model others do not know yet. Let's make the introduction. What do you think?"

"Let's start with the first model," Eohn says. "Everything around us is projected as different images on a cube's faces, or on six-sided dice. On surfaces, are six different numbers, or frames. On the first face, we see an image. As an example: we observe a forest.

On the second face, we hear sounds—leaves moving, water dripping, wolves howling.

On the third face, we smell—a forest, flowers, the earth, fir, lime, water.

On the fourth face, we feel tree bark, grass, leaves, spiderwebs, water, and so on.

On the fifth face, we taste—blackberries, apples, hazelnuts.

"We have linked each of our senses to one side of the cube. With these multiple perceptions, we recognize nature, and it enters our memory. In this way, we make a complete picture of our environment through our senses. We cannot be deceived."

"There is one last face," Reader says. "We can relate it to another function, another perception. We can associate it with intuition, reason, memory, and knowledge."

"Let's have only five faces," Ahve says, "a pyramid and its base."

"The pyramid can symbolize the five senses through which we know the world," Reader says.

"We're not going that far. Let's go back," Ahve says. "What schools teach children only through their five senses? Maybe only those in jungles rely solely on the five senses. Through education, we can develop the future society. All human and material resources must be used here without any restrictions."

"Let's look at another pattern," Eohn says. "When we were born, we weighted perhaps three or four kilos, but at maturity, we weigh perhaps seventy kilos. Nature has fed us. I grew up based on what I ate and drank. Nature is our second mother. Would her children kill her? That would be a horrible crime. What is in nature is in our bodies. If we poison the air, water, and earth, that poisons us and other creatures. We're doing wrong, and it comes back to us. We should follow the golden rule in this regard. There is a saying: "What you do not like to others does not do.."

"Past evil repeats itself," Ahve says, "in very subtle and diversified ways. Dictatorships, torture, espionage, and lying in many forms. Let's consider slavery. Slaves were property of masters. They possessed nothing. They had to work, were kept in chains, and had to labor for their masters for free. Slavery has been abolished, but the attitude and behavior remains—we are enslaved but very subtly by technology, the media, and propaganda."

"How can you say that?" Reader says. "People today are not in chains or killed by masters."

"Those who are arrested by police are put in chains," Eohn says. "You have to see people who are financially chained through the debt they have to repay for the car, home, health, and more. To obtain the necessary money, they work for the employer; the part of his best life consumes from 9h to 17h. Like the animals are yoke by human beings, today countries go to war to enslave each other, as from antiquity did."

"Prosperous ancient Athens was based on slave labor," Reader says. "It developed commerce, manufacturing, arts, and science and underwent tremendous development, but it was based on slave labor. Nonetheless, it developed the concept of democracy. We can compare slavery then with slavery now."

"We know that slaves in Athens had access to all activity but political activity," Ahve says. "Who has this status today?"

"Those today younger than eighteen cannot vote," Reader says. "They are the equivalent of slaves."

"That's right," Ahve says. "Slaves worked in fields, mines, workshops, and homes for domestic purposes—cooking, cleaning, and raising children among them. They were considered members of the family. They were entrusted with money for shopping, and they had the right to wear a short sword for defense. They could buy and sell, trade, and own property with the consent of their masters. They could also buy their freedom, and they could participate in war. At the Battle of Marathon, over 15,000 slaves fought on the side of the Greeks. After that service, they were freed. Athens' police force was made up of slaves."

"In Athens," Eohn says, "slaves were people who had to obey their masters, but later, in the West, they were property much like animals even in Christian countries. That is so degrading to our history."

"Do we now have analogies to slaves?" Reader asks.

"As I said before, what was, there will be," Ahve says. "Vices are passed on. Events are cyclic."

"Madah, I feel you're against money, finance," Reader says. "How do we promote democracy, freedom?"

"All means used in moderation can benefit everyone," I say. "Money used with temperance can be helpful, but we have financial tyranny today. Money has become a weapon used against people, nature—for instance, high-interest loans that are hard to pay back. People who came to power hijack it for their ambitions and manage to destroy human and natural resources. We criticize that to stop it from repeating. Everything you said serves to preserve life: people, fauna, and flora."

"I was thinking only of myself," Reader says. "I forgot that there are other beings on earth. Through them, humankind has developed, and through them, we will expand into the cosmos."

TYRANNY AND DICTATORSHIP

"From the old days the people were abused by leadership," Eohn says. "That's what happened to religion. Consider, the Inquisition. From an extraordinary biblical history: Jesus Christ has healed people and, raised the dead. On the cross with terrible pain gives by the nails in His body, He said: 'Lord forgive them for not knowing what they are doing.' He rose from the dead and went to heaven. By His death and return, He gave us life and hope, but the pope and the cardinals gathered armies, conquered territories, and spread death. The Inquisition burned people alive and arrested and tortured innocent people. They did not follow the Savior's teaching. The dictatorship model of the Inquisition has been copied all the way to the present. We have dictator states, dictator finances, and dictator institutions. Dictators grab and abuse power with much cruelty and destroy people and nature."

"Do you think this is happening today?" Reader asks.

"Old habits have been preserved," Ahve says. "Today, they are more subtle, but they still divide and rule to make people poor and docile. In cities, people work at different jobs. They do not farm or raise animals. These means of subsistence were taken away long ago; without them, they are easy to handle and trick. Dissidents aren't killed. They are subjected to a more subtle social death. They aren't allowed to work, so they cannot support themselves. The socially dead are thus eliminated."

"This means that cities are camps for millions of people," Reader says. "Their freedom is staggered by the money they possess. The attractions of living in cities are schools, hospitals, libraries, institutions, shops, and workplaces. Let's be careful what we criticize."

"One of the pillars of Western capitalism is private property," Eohn says. "The owner of a business or farm has unlimited power. A state constituted based on private property cannot be democratic. Dictatorships in organizations are transferred at the state level. Leaders exercise absolute, totalitarian rule over human and natural resources. The forms of control and possession are more subtle, but they are powerful. What are we working on to change this?"

"Certainly, we were talking about the notion of property," Reader says. "Power over the property should be limited, be not absolutized. We divide the issue into only two parts. I think about the time duration of the power property that has to be limited and the level of power.

Absolute power has to be limited to "partial power" - meaning, that an institution should not make products for killing or products from slaughter being.

"Ownership time" has to be limited to five or ten years, so the chance for abuse is reduced."

Abusive capitalist practices oblige consumers to pay several times for what they buy.

"Bandits extort money from small businesses and burn them down if they don't pay continually. You have to pay for some software programs yearly. That practice has the consent of the state, which takes its share in taxes. My computer programs are not mine; they can be taken away if I don't pay yearly. The state can manipulate and rob people through the internet. Do you agree with that? Is this a dictatorship? Are we serfs?

"Computers can take us hostage just as people were taken hostage in ancient times for ransom. When I go on the internet, I can be stuck as ads are displayed. I'm taken hostage; I'm immobilized. Some of my life is lost.

"Other practices also violate human rights, individual freedom, and the right to property. These happen in states that call themselves democratic. We are not in Congo or Mexico. Such actions kill my time as they do others' time. Free countries outlaw taking hostages, but they still do so, and nobody protests this. Why not?"

"But companies pay for commercials," Eohn says.

"I'm held hostage, however. I am robbed by these practices," Reader says. "I'm paying. Why doesn't my money have the same value as that of companies that advertise? The confiscation of individual property by companies is abusive. We should have a referendum to nationalize companies."

"This is a critique of the current system based on catastrophic exploitation of people and nature," Ahve says.

"If large private enterprises can be expropriated, can new ones be established?" Eohn asks.

"Yes," Reader says. "People can invent things and control the means of production but for a limited time, say, five to ten years. After that, it becomes public property."

THE PROBLEM WITH MONEY

"Money is being raised to the rank of divinity," Eohn says. "To abolish money, we abolish the notion of property and make access to air, water, food, education, health, and other necessities free. Resources can be plentiful for all if we can achieve unlimited energy."

"But society is currently using money everywhere, correct?" Reader says.

"No. When you were little, your parents fed, clothed, and sheltered you but gave you no money," Eohn says. "That practice should continue to and after adulthood. It's a natural process. In so-called primitive societies, this continuity exists. Money divides society into two: the rich and the poor, who are exploited by the rich. The wealthy manipulate and subordinate people. They want to be remembered for their monumental achievements—pyramids, castles, buildings, art, and so on. They can have the same or greater satisfaction by preserving life on earth. Money is a weapon for maintaining injustice."

NATURE AND MONEY

"Money is misappropriated," Eohn says. "It is an artificial product made by man like houses, cars, bombs, explosives, and other means of destruction. Money as an artificial product can be exchanged for other man-made products such as bicycles, shoes, and so on. We want to prevent living things from being sold and bought. Nature does not make money. Nature does not recognize money as a medium of exchange. Extraterrestrial civilizations do not value money; they see human foolishness in it.

Can you tell me how much it costs to cut down ten square meters of a rainforest or take a ton of fish from the ocean?"

"Let's say $10," Reader says.

"So a hundred square meters of deforestation or ten tons of fish would cost $100?" Eohn asks.

"That money goes to pay wage, pay for machinery and fuel and so on," Reader says. "But nature receives nothing. That's not fair." This money gives them a salary to workers, pay for machines, fuel, and more. They do not get into nature. Nature receives nothing. It is not fair.

"The money workers receive goes back to the capitalists," Eohn says. "They have to pay for their housing and food, so they end up with almost nothing; the poor remain poor, and the forest and oceans receive nothing for the wood and fish. Do you see the similarity? And forests are not replanted, and fish are being harvested to extinction. As you can see, giant trees are cut, these natural monuments are destroyed with animals, birds, insects, the surrounding flora. It's a massacre. They cleanse everything. They not leave a portion for restoration.

They do not have the common sense to plant seedlings, shrubs for reconstruction. They do not have the wisdom to supply small fish, to restore the captured fish. These beings have no representatives in parliament, in Congress. Where is justice, fairness? How nature is exploited, so are people being exploited."

"Can we replace money with another medium of exchange, storage, value?" Reader asks. "I would give the example of energy."

"We have tried to show how behavior toward nature is reflected in society's behavior," Eohn says. "The evil practices of the past are being repeated. I urge you to find other vices to be removed from future civilizations. Money used to be gold, silver, copper, and nickel, but it has been reduced to numbers stored on a computer. Old forms of money are on the verge of extinction even in capitalism. They operate with these vices that are hidden, controlled, speculative, and profitable for banks. It has come to the paradox of money making money." The best thing to do is to have a good education and treat others ethically and with compassion. There is no point in looking for a substitute; in more-advanced concepts, money exchange has no application.

"Because money is an artificial product, it should not be used by to buy a living being in nature. Man cannot make even a bacterium much less a plant or animal. Sales and purchases of living beings are injustices and must be banned, treated as a criminal act against life. The above brief critique serves as a judgment to take the best of human culture into the future—art, science, medicine, industry, agriculture, and others. We have tried to show how behavior toward nature is reflected in society's behavior. Old, evil practices are being repeated. I urge you to find some other vices to be removed from future civilization."

"This reproach does not seem consistent to me," Reader says. "I have an idea: strong states have agreed through treaties to reduce nuclear weapons. Let's also propose a money treaty to reduce their vices and develop their virtues."

CHAPTER 8

PREDICTIONS

"We imagine applications by using the notions we have developed. We want to develop them for life, people, and other living creatures who are our cousins on earth," I say. "I take into consideration the following criteria. Society goes through changes. It can choose decay or life, and I think it will choose life. We have fast means of transport, enough energy, and depollution. We should make preserving life our highest priority."

TRANSPORT VEHICLES

"At present, transport is based exclusively on the law of action and reaction. We move to the next level of development based on the inertial-force generator. This is an extraordinary leap—a thousand times more effective than at present."

VEHICLES WITH CONSTANT SPEED

As a reference, table 2 offers data for vehicle ratings.

Power tables. For force F = 40,000N

Nr	k^2	$(1-k^2)$	P = Ec/T [W]	F/P [N/W]
1	0.9	0.1	200,000	0.2
2	0.95	0.05	100,000	0.4

3	0.98	0.02	40,000	1
4	0.99	0.01	20,000	2
5	0.999	0.001	2000	20

Table 2

"We have two cases that we are studying: with constant speed or constant acceleration."

"What's the difference?" Reader asks.

"The constant-speed concept is used for small vehicles comparable to a bicycle or motorcycle," Ahve says. "Travel over short distances, ten kilometers to a hundred kilometers, takes ten minutes to a hundred minutes. Travel speed is this.

10m/s (36km/h) … 20m/s (72km/h)

"Is this like a flying bike? A toy?" Reader asks.

"It's a lightweight, easy-to-use sporting vehicle for young or old," Ahve says.

"Consider the example," I say, "of the children going to school. The team has ten children; five children eight years passive and five children active fourteen years of age that can generate 200 W each; in total the available power is: 5 * 200 = 1000W.

From table 2

$k^2 = 0.98$; F/P = 1 N/W

"we get a force

Ftraction1 = 1 * 1000 = 1000N = 100kgf

"The mass of passengers is about

10 * 50kg/body ≈ 500kg

We evaluate the vehicle's mass, 500kg. In total we have: 500+500=1000kg, passengers plus vehicle.

"With a force

Ftractiune = 1000N

"the device can fly at an average speed of 72km/h.

"For $k^2 = 0.999$,

F/P = 20N/W, Ftraction = 20 * 5 * 200 = 20000N = 2000kgf

"It appears that in this case, the vehicle can take off, land vertically like a helicopter, and engage in horizontal flight at a speed of up to 100km/h."

"Why limit speed?" Reader asks.

"Movement is made to eyesight," Eohn says. "Human reflexes are not very fast, so we limit speed. In these slow machines, it is not necessary to provide them with complicated equipment. They will be used by children and adults. The distance depends on how long the human engine resists. For safety, we have glasses, gloves, seat belts, and parachutes in case of danger so that the vehicle comes to the ground at low speed.

"There is also an interesting discussion about how we use human energy: like a bicycle using the legs or like the rowing boat that uses the whole body. I opt for a movement that uses the whole body that achieves the harmonious development of the people."

"We can make vehicles move on the earth and water, and even in the cosmos," Reader says.

"Of course," I say. "We may move on the ground or in or underwater slowly, and in space, we are limited by the reserves we carry. But we have two extraordinary achievements: obtaining traction by inertial forces and recovering linear energy and rotation. These two will propel us fantastically, and we can by our imaginations conceive other machines."

VEHICLES WITH CONSTANT ACCELERATION

"It is the engine of humankind through which it can have achieve the unbelievable. Consider a vehicle that has the mass plus the 10000kg load.

"For $k^2 = 0.99$,

F/P = 2N/W,

F traction = 200000N

"for take-off with a = 10m/s^2

"In the cosmos it can have,

a = 20m/s^2.

P = 200000/2 = 100 000W = 100kW

"This is the engine power to compensate for losses.

"We want time to go for a distance S; we start with the formula: S = a * (Ta) 2/2 + a * (Td) 2/2; Ta = Td;

Where: S = distance traveled [m], a = acceleration, deceleration; Ta = acceleration time, Td = deceleration time Ta = Td; T = Ta + Td ; total time.

S = a * (Ta) 2/2 + a * (Td) 2/2; Ta = Td

With this, the travel time, T will be

$$T = \pm\sqrt{\frac{4S}{a}}\,[sec]$$

"For example,

$$S = 100km = 100000m = 105m; a = 10m/s^2; T = \sqrt{\frac{4*100000}{10}} = 200s = 3min20s$$

"The maximum speed is for:

T/2; Vmax = 200s/2 * 10m/s^2 = 1000m/s

"if we consider that it also rises vertically, H is height, altitude

H = 10 * 100^2/2 = 50000m = 50km

"You see that transport times are much lower than the current means of transport. We have the means to move to the planets in our solar system. We can satisfy our curiosity to investigate what is in space. It is very important that we have huge spaces for agriculture, industry, and others. No more need to aggressively exploit earth for human expansion."

"So we have ample space to use, short flight times to get where we want in the cosmos," Reader says. "I think we can play with time. Maybe we can travel to the future or the past."

"Out of curiosity to see what is the travel time for different distances, the space crossed, S."

	S [km]	S [m]	a[m/s^2]	$T[sec]; T = \sqrt{\frac{4S}{a}}$	T [s; min; h]	Note
1	100	10^5	10	200	3m20s	
2	500	5 * 10^5	10	447	7m30s	
3	1000	10^6	10	632	10m32s	
4	5000	5 * 10^6	10	1414	23m34s	
5	20000	2 * 10^7	10	2828	47m8s	Half the equator
6	~350000	3.5 * 10^8	10	11832	3h16m	Terra to Moon

7	-60 * 10^6	6 * 10^{10}	20	109544	30h25m	Terra to Mars

Table 8.1 Driving times for vehicles with a = constant.

"In my opinion, this is fantasy," Ahve says. "We cannot go back in time. We might want to give the planets and galaxies back to reverse the biological processes in the whole universe, but we do not have the means and knowledge to do that. "From the time formula

$$T = \pm\sqrt{4S / a}$$

"We have a positive value and a negative value. The negative value is mathematically correct, but it has no real physical meaning. To give time back does not seem possible.

Let example with a positive time value for a thousand kilometers. The trip takes sixty minutes by a plane or six hundred minutes by car, but only ten minutes through the concept of acceleration and deceleration after by physical laws. "

"But if our vehicle accelerates for a year or two, could we overcome the speed of light?" Reader asks.

"Of course," Eohn says. "We can achieve any speed by accelerating by our own resources not relative to the environment. Thermally or electrically produced light has constant speed; it has nothing to do with our vehicle. The danger is in clashing with matter in the way. If it is far superior to the speed of light, it is possible to pass it through other bodies without feeling it. It's like being in another dimension. It can traverse planets, suns, comets ..."

"You contradict Einstein," Reader says. "He says that the speed of light cannot be exceeded."

"He is right if the ship is propelled by a beam sent from the earth," Ahve says. "We imagine that we are in an area where we do not see anything around. We can accelerate indefinitely. So we can achieve any speed we want. The problem with speed is that we need a fixed point of reference, that does not exist. There are interesting things we do not talk about now."

"What do we use these vehicles for?" Reader asks.

"For passenger transport," Ahve says. "We could take twenty people a thousand kilometers in ten minutes for very little cost. We will not need airports and all their facilities."

"We can transport goods, animals, and plants anywhere," Reader says.

We can send where it takes: water, food, medicine.

Eohn: "One of the most remarkable applications is personal transport," Eohn says. "By air, we can go wherever we want without being stuck in traffic, at stops, or at intersections, so we will no longer need roads, bridges, or tunnels. We can spend for other human needs from what we save money remaining from unnecessary roads superstructures. "Our firefighting vehicles and ambulances will get where they need to go quickly," Ahve says.

"We can use as tractors to lift weights as like cranes do now," Reader says.

"Transport to and from the cosmos is important," Eohn says. "We can take people to space colonies along with water, earth, plants, and machinery and come back with products made there."

"That would be a great achievement," Reader says, "but it seems impossible. Why have space agencies with scientific, economic, and military interests not done so already?"

"Maybe they have such knowledge," Eohn says, "but scientists reject the idea that a vehicle can be propelled by its own means. It is a prohibited subject. Those who suggest it are labeled as stupid, but we have shown that it is possible. More-advanced civilizations closer to us, say, ten light-years away, have done so. Their UFOs are a source of inspiration that can urge us to unravel the problems and issues involved. It's an invitation for us to advance. They call us to join them. Do you think they're making friends with those who kill life? Would you do that?

"We still have the assumption that they are civilizations biillions of years old that have knowledge that exceeds ours. In legends, religions describe events that induce the hypothesis that they have intervened in the history of humankind.

How do you understand God? I suggest that it would be a civilization that is billions of years old."

GENERATING ENERGY

"Another very important application is getting the energy we want," I say. "We do it for people and for all the creatures. A satisfactory value for a collectivity is

1MW = 1000kW = 1000000W; we can choose lower or higher values. Let's make a design of it starting with * figure 4.1.

"Choose:

F = 100000N, v = 10m/s, r = 10m, Power compensation = 100000N/2N/W = 50000W; for : $k^2 = 0.99$

"The power generated is

Pg = F * v = 100000N * 10m/s = 1000000W

"The available power is

Pd = 1000000W-50000W = 950000W ~ 1MW

"That's enough power to provide for over a thousand homes. And because it is generated locally, we do not need long power lines. It could be used on land or in space stations. We would not need hydropower or thermal power plants or nuclear. We could create independent, comfortable energy. Now let's see what we can say about the future."

HOUSING

"Our homes are thousand-year-old concepts," Ahve says.

"But we now have introduced heating, air conditioning, windows, security doors, electric lighting, elevator, television, water, sewage, and other developments," Reader says.

"Out of air conditioning and electricity, comfort in luxurious homes existed a long time ago.

"Let's formulate a concept closer to type-two civilizations. What is a house used for there?"

"Eating, sleeping, storage, protection from the environment, and so on," Reader says.

"We are optimizing life through available knowledge," Ahve says. "Let's consider sleep. Why do we need to sleep?"

"The sleeping goal is to regenerate your body, to balance the body, to feel vigorous when we wake up." Reader says

"Let's say that optimal restoration is in the mother's belly," Ahve says. "Optimal temperature, humidity, and power supply. We transfer these conditions to the sleeping man."

"We perceive temperature in three ways," Eohn says, "as heat by convection, radiation, and direct contact. Bedrooms are optimal places for rest. Radiated heat is generated by electrically heated walls or by the circulation of the hot liquid inside the walls a fluid or by the circulation of hot gas through the walls. Air temperature and humidity are adjustable. We may purify the air, mechanical, electromagnetic and by adding negative ions we increase immunity. The air may be enriched with natural stimulating essences such as oil from flowers. We can also look for other regenerative stimuli, energizers.. We appreciate that vibrations, mechanical movements can be useful.

"We consider the optimal state is that in which the body consumes the minimum energy for balance with the environment. Consequently, the majority of the body's resources are directed to regeneration, recovery."

"But that costs a lot," Reader says.

"We do not consider money," Ahve says. "The quality of life is so much more important. We should conceive of everything possible to prolong life, increase performance, and ensure the health of people and other creatures. We should use all technology for these purposes. Beds are made of lightweight metal with carbon fiber mesh. Electric conductors discharge electrostatic charges, and carbon nets can be connected to an electrical source for other applications."

"Can we use sleeping quarters for other activities? Reading? Listening to music with friends?" Reader asks.

"Of course," Eohn says. "The concept is for colonies anywhere on earth or the cosmos. We can design spaces with specialized, optimal facilities for all human activity. You will not think of money, which kills living beings and destroys nature."

"There is still much to say and do," Ahve says. "Let's first take the concept of disease prevention so we will not need doctors."

"How did you come up with that idea?" Eohn asks.

"I'll tell you," I say. "When I visit my doctor, he tells me, 'Take these pills. Goodbye.' When I was younger, my doctor would check out my throat, ears, and eyes. He would push on my liver, look at my joints, listen to my heart and lungs, and so on.. Physicians in the West today are afraid of contact with their patients. So we say we do not need them. We can replace them with machines that do the same thing."

"We can do permanent health monitoring for each individual," Ahve says. "We can conduct a microbial and chemical analysis of the water they use to wash their hands and face and their urine, feces, and saliva—whatever leaves their bodies, and that includes their hair and fingernails. Infrared photography of their bodies will show abnormal temperature distributions. Internal body views will investigate the condition of their joints, skeletons, blood vessels, hearts, kidneys, intestines, brains, and so on. We can make videos of their posture and movement. We can collect other data as well, and these can be very important for children, who are always developing. All this data goes into a medical dossier, and computers can analyze it rather than doctors doing that and identify abnormalities and deviations from the norm. This is briefly the preventive part we are developing."

"Such analysis will require a lot of equipment and large spaces for it even if we do not have a problem with how much it costs," Reader says.

"In the future, all of this will be handled by a computer no larger than a cell phone," I say, "that connects to sensors, receivers, image converters, and other devices necessary. What we do for humans is being done for other beings in a lesser form already."

"But some machines make such analyses already," Reader says.

"We can make them even smaller," Ahve says.

"This way, we can have preventive monitoring of human beings and fauna," Reader says, "but what do we do in the event of accidents or severe illnesses?"

First, we have hospitals no farther away than a thousand kilometers from each other," Ahve says, "and we provide them with all possible means of medical intervention. Fast ambulances can handle the transportation of patients to these hospitals for diagnoses and treatment including surgery. For flora and fauna, we can go to nature or bring them to the hospital."

"Why heal the nature?" Reader asks. "Why try to help lions, crocodiles, antelopes, zebras, elephants, and trees or diseased plants? Carnivores prey on herbivores, which eat vegetation."

"We have to ethically keep all beings alive," Eohn says. "They all contribute to life on earth in their own ways. Without predators, herbivores would destroy vegetation. Carnivores keep herbivores in check. In that way, carnivores protect vegetation. Altogether, they form a balanced system of life."

"We people, supposedly intelligent beings, kill without regard to this balance," Eohn says. "We can see in grocery stores so much blood and meat. Predators have the wisdom not to exterminate the animals that are their food; they eat as much as they need, no more. People, however, do the opposite. In Europe, they destroyed wolves and other animals. In America, they destroyed the bison. In Asia, they almost exterminated tigers, and in Africa, they reduced if not destroyed lions, elephants, and monkeys. Buddhist monks protect and nourish tigers; they provide them with comfortable lives. They walk these big cats in parks among people without incident. It is remarkable the love that surrounds them. Raised as kittens, they become just big cats that like to be comforted. In the West, we lock up animals in cruel zoos for profit and entertainment, but some people care about and love such animals."

"We have deviated from the subject," Reader says. "What can we say about conditions for humans in space colonies?"

"It's time to talk about lifestyle changes," Eohn says. "Lifestyles change with time and technology. I'm older than you. When I was a kid, I played outside with other kids, climbed trees, bathed in the river, and took goats out to graze. Now, I see my grandchildren sitting all day playing computer games and surfing the internet. They and even their parents don't go outside; they remain motionless. We should offer more physical activity for their harmonious development. We say that scientific progress, technology, automation, and robots make work easier and increase productivity, but it has negative effects in that it removes people from nature and physical activity. How do we solve this?"

"We could promote vigorous physical activity with flight machines powered by the force of people's muscles," Ahve says. "We could exercise on our way to or from work for thirty, sixty, or ninety minutes. We could promote a mass movement that gets all citizens involved in sports or other such physical activity. And we still have a powerful tool—inertial exercise that promotes harmonious physical development to counteract the effects of all the time we spend motionless."

"As we advance due to computers, cars, and other commodities, that will remove human beings from nature," Eohn says. "What do we do about flora and fauna?"

"We transfer technology to nature," Ahve says. "We teach beings to activate mechanisms to give them water, food, and shelter. Living beings, including vegetation, are intelligent and capable of stunning action."

"This is how we could contribute to the development of Terra's flora and fauna intelligence," Reader says. "Perhaps we could also contribute to the development of consciousness in all creatures. We can develop a kind of school for them. We have to preserve their natural environments. We can experiment at first with dogs, cats, parrots, and fish for instance. Exciting experiments await us."

"In large colonies, we can organize restaurants for whole populations—toilets, sports fields, training rooms, and other facilities," Eohn says.

"I still need to briefly discuss the life in these collectives," Ahve says. "The magic number is three. We have children, parents, and grandparents. Parents go to work, and grandparents take care of the children. They feed and dress them, tell them stories, and pass down traditions and customs. They teach agriculture and animal husbandry to introduce them to nature. Grandparents actively participate in the production of food for the colony. Through such preventative and regenerative methods, third-age people will be vigorous and healthy.

"Can we currently do something to get closer to the future?" Reader says.

Of course, an accessible experiment for everyone curious about life in space would be to experience the cultivation of plants, vegetables, fruits." Ahve says.

Let start to grow tomatoes, we plant seeds in pots and give them measured amounts of light, water, and warmth, and we analyze the results in terms of harvests and taste. We study the productivity of the soil to determine optimal growth conditions and share the results with everyone. Like that idea, we may find others."

CHAPTER 9

ASSUMPTIONS ABOUT MORE-ADVANCED CIVILIZATIONS

"Can you tell us more about more-evolved civilizations?" Reader asks.

"I have no specific information you can rely on," I say, "so it is difficult to advance in this direction, but we can use our imagination. We can say that the ease of traveling in space presented above is an argument that more-advanced intelligent beings might have it as a starting point. We can try to find out how capable they are even if we exaggerate."

"I choose a reference source—legends, myths, religious writings, verbally translated folklore, interpretations of cave pictures, and more," Ahve says.

"But those are just stories, fantasies," Eohn says. "If they started from facts, it will be difficult to separate that from the metaphorical, poetic additions. We can eliminate from these stories what seems to be exaggerated."

"If we consider only what we see as plausible, we could lose important things," I say. "I believe we should take the original descriptions and give them possible explanations.

Let's start with the concept of God. Does God exist or not? God is infinite, eternal, omnipresent, and omnipotent. I personally do not understand him; he doesn't fit in my head. He is a very complex concept. I do not have him in me as a mental representation or like faith. I like to believe in Him if I were convinced that it exists.. For me, it's easier to understand that angels are intermediaries between us and God. They resemble us. They are strong, and they perform the wondrous. I perceive them like extraterrestrials being."

"You know God by faith," Eohn says. "Through prayer, you can get close to Him. He gave us immortals souls that animate our mortal bodies. He urges us to do good. We can read the holy books and pray so we can feel his infinite power."

"I have had patients who after clinic death said they had detached from their bodies and seen themselves from above," Ahve says. "That would be their immortal souls, their astral bodies. They met the souls of their departed loved ones. They saw a white light, and they say they met Jesus."

"We need to make a break," Eohn says. "I think such phenomena—UFOs, extracorporeal evasions, and Buddhist practices are widespread in order to move us away from our ancient religion. They want to prepare the ground for a substitute, another religion. The goal would be to subordinate us more strongly."

"Most doctors say it's a hallucination of an oxygen-deprived brain," Ahve says. "Others claim that it is a real phenomenon worth studying. We do not deal with it even though it's an interesting subject. I think it's a hallucination. We are examining the interval between birth and death.. We do not know what happens before birth and after death. We can't experience that. We consider this a secondary concern; we let it mature."

"People with all that future comforts will lull, degrade, and perish," Reader says.

"Let's move on," I say. "Many peoples have in their legends the belief that the gods who came from heaven have given them agricultural knowledge, metalworking, and astronomy. They promised to come back. It is an indication that there have been interventions from beyond earth."

"Those are just stories," Reader says. "People are smart enough to come up with these things themselves."

"Does anyone have another explanation?" Ahve asks.

"Let's take the Bible as the source," Eohn says. "It contains wonders. How did Jesus turn water into wine? There is a hypothesis that organisms can undergo biotransmutation—a chemical element can become another element. Some chickens can convert potassium into calcium by adding a hydrogen nucleus. The reaction takes place at low energy levels.

"If bacteria, plants, animals, and people can do that, aliens believe they have good control of the phenomenon. Jesus had the servants pour water into vessels at the wedding in Cana and transmuted it into the chemicals that constitute wine: alcohols, acids, esters, aldehydes, carbohydrates, minerals, and vitamins. This transformation shows that they have very advanced knowledge."

"We have also said that there are extraterrestrials with differentiated development levels," I say. "Not everyone is the same. I think Jesus was far advanced. I'm thinking of his multiplication of the bread and fish. I suspect he transformed atomically, chemical and biological structures to do that as well."

"Consider Lazarus's resurrection four days after his death—his rotten, malodorous body was returned to life," Ahve says. "Besides transmutation, they rebuilt chemical structures and restructured nerves, blood, and cells. For many, that was a miracle, but through our imagination, we can explain what might seem impossible."

"There is also the resurrection of Jesus to consider," Eohn says. "He died but rose from the dead. Here, I see that unseen external forces have intervened—in short, a miracle."

"You give fanciful interpretations of events that are imaginary," Reader says. "Religions are useful to those who lead. The idea of a reward after death tranquilizes people. They thus bear the humiliations and sufferings to which they are subjected. What can we do?"

"Should we try to unite through the power of thought?" Eohn asks. "As I suggested, we could all think of a common action for three minutes at a certain time, say at noon. The union of thought will be followed by association between people and action for life on earth."

"In our lives, we have shown that we can have fast transport into the cosmos and the earth, unlimited energy, and symbiosis with all beings on earth," Ahve says. "We can try to develop ideas for optimal life and evolution for all living creatures. There is no science, no philosophy, no concept that offers such a perspective. Do you know an alternative?"

"I do not know if there is one," Reader says. "Are scientific people, politicians, leaders, that talk to us. What we are perceiving as is a mix of propositions that we don't understand. I want to be with you, but I have my doubts."

"Try to find your own way. Try to find other explanations," Ahve says.

"I'm like a grain of sand among all humanity," Reader says. "We need a leader to show us the right way. Perhaps in time, based on what we said, we can recreate our concepts and ideas. There are people with special facilities to do this synthesis. I think the Bible was inspired by aliens. Through Revelation, it predicts a change of humankind. They said there would be false prophets who would deceive many. What have we said so far is a prophecy?"

"The Bible deals with the relationship between humanity and God," Ahve says. "We deal with the relationship between humans and aliens; we follow another path. We want humankind to advance toward them, to become worthy of being virtuous intelligent beings. We want the good of all the creatures on earth. What I have said is a possible prediction of the future. A vision is not a prophecy."

CHAPTER 10

SOURCES OF INSPIRATION

"Madah, tell us if these discoveries belong to you or if someone gave them to you," Reader says. "Did you have unusual inspiration?"

"That's why I've been working for over fifty years," I say. "I think what I have discovered does not belong completely to me. I think I was subtly guided by other forces; this doubtful assertion is based on my intuition."

"Even if they are vague impressions, tell us about them. We can learn something," Reader says.

"Before I was born, my mother had a child who died at age two," I say. "She could not have another child though she wanted one. She was very faithful in praying for another baby. After a long time, she dreamed that St. Peter and God appeared to her. She crossed a bridge, and they said, 'Woman, you will have a baby.' That is how I came to be. Maybe I have premonitory sensitivity.

"I spent three years in Bessarabia with my mother. I wanted to go back to Timisoara in Romania. I saw on some frosted glass images of buildings, houses, and trams that stayed with me. I was certain we would return. I was fascinated by the gyroscope as a twisted rope and others.

"At university, I had a professor in mechanics, Mihai Stoenescu. He told us that research is of two kinds: one that starts from the existing formulas that it processes and develops, and the second, which starts from experiments—making a physical model, a mathematical and experimental model. The second type of research seemed more interesting and adventurous to me.

"I have been interested in UFOs since my youth. I said that they used physical laws to travel that we humans could discover, and I said that I would work hard until retirement to discover them. Eventually, I found something after making tens of thousands of calculations. We have advanced extremely slowly. Every time I learned something, I developed calculations that gave positive results and then performed experiments to validate them. I have discussed the essence of what I have discovered with you."

"You have something unusual that helped, favored, or prevented you from doing this research," Reader says.

"We have native languages that we speak from childhood," I say. "The structure of the language forms the mind, the brain, and our behavior toward nature and people."

"Some young children who were lost in the woods survived because they were protected by wolves, monkeys, goats or other animals," Ahve says. "After several years, they were found. They held onto the behavior they learned in their natural environments. If they had been raised by wolves, they behaved like wolves. The brain is programmed in the first five years. So today's children trained from kindergarten on become easy to handle. They are programmed to be children of the city. They fear mice and spiders and do not know Terra. They live in front of TVs, computers, and cell phones. They are programmed to stay in closed spaces, and they become dependent on what is found in stores, artificial products. They work for bosses, pay taxes, and obey authorities. They assimilate ideas promoted by TV, press, advertisements, and propaganda. The environment and language structure brains and thus their reaction to events."

"Thank you," I said. "When we speak, our words have power. My native language is Romanian; it has a syntax that is reverse of those of Western languages—French, English, German, Italian, and others. We say, 'I see a country that is rich,' in Romanian, while in English, we say, 'I see a rich country.' In Romanian, the syntax puts the subject first; the noun, the object, the work, and the philosophical category that is concerned that generates the image of territories with mountains, forests, plains, animals, and people. The second is the adjective—the qualities of the noun.

"In Western languages, the appealing qualities, the adjectives, are emphasized for the first time. The variety of wealth stimulates the desire to possess it; it induces an aggressive character to those people to take that wealth. The noun follows it and shows who belongs to those qualities. Thus, through Romanian, I focus on the essence of phenomena, not on their quality. Representation starting from the noun, the subject, gives a different way of conducting research. I had the advantage of a language that favored my achieving positive results."

"This means learning Romanian opened up other visions for you," Reader says.

"You have to teach it from childhood," I say. "You are thinking through it to perceive with all your being. If you learn it as entertainment at age thirty, you do not get to its essence. Talk to Romanians and you will discover that they have opinions that would find very different than English, French, speakers. Also Russian is flexible."

"From the past, I did remember very vague pressure feelings. I feel that forces outside me induced in my mind feelings of gratitude when I was doing what they asked I was asking or sadness and depression when I was not working for them. I lost some jobs because of that. After several hours of work, I was led, pushed to think of these propulsions I had not escaped. It became evident that my productivity was down."

"Are these propulsions dangerous?" Reader asks.

"Yes," I say. "Again, through them, we can change the mass center to the planetary level. We can make the earth and the sun draw closer or move farther apart. We can change their rotation and make the day longer or shorter. Other geoclimatic changes may occur. People must restrain themselves from changes that may have unfavorable consequences."

"These achievements can lead some people to use them to dominate others and exterminate life," Reader says.

"That is a great danger," I say. "That has occurred before. Just as nuclear power can be used for good or evil, my ideas about propulsion can be used for good or evil."

"Do you think there may be vehicles with superior performance over what we have presented?" Reader asks.

"I see that UFOs have unequaled performance," I say. "These mass accelerations are different according to their level of progress. Concerning mass acceleration, there are other more-evolved possibilities that belong to super civilizations that are above our understanding.

"We are at a lower level. We have concentrated forces at the center mass point. Starting from this concept, we have to advance several levels or steps. We need human, material, and time resources. Every step of progress requires fifty or seventy years, so we will have to rely on later generations."

"From the beginning, you were possessed with the desire to learn the truth. Did you find it?" Reader asks.

"I have advanced in that direction," I say. "We see that reality can be viewed from two points of view. We have the dice symbol, the six-sided cube; we stayed on one or the other in succession. First, we see from the perspective of human culture how science, technology, and trade are at the service of the financial, industrial, and state powers. They aim at controlling and exploiting the environment and people. The concepts of concurrence, competition, free market, profit, and conflagration are used. These are visions of war that lead to the destruction of life on earth. Current human activity leads down a road from where there is no return.

"The second point of view is that we place ourselves in an advanced, type-two civilization. The vision here is in great contrast to the previous one. We have access to fast transport on earth and in the cosmos, unlimited energy, large spaces, and food resources for humans and all living beings. Their work is directed toward maintaining, developing, and spreading life. It is good to know both. Reflect and combine to discover other aspects Choose which truth suits you best. Other truths will appear."

"Great achievements and developments are said to belong to large corporations and institutions," Reader says. "One person cannot accomplish very complicated things."

"On the contrary," I say. "An idea starts with one person. Organizations support it with human and material resources to materialize it in hopes of making a profit. We say that achievements such as planes and computers are complex, but they began with one person."

"You're describing an unrealistic utopia," Reader says.

"It is a real description of a bright future," I say. "The scientific part involves calculations and explanations. There is still much to be done. Experiments on a small scale produced positive results, and on that basis, we develop a vision of the future we want. The predictions made can be the source for your developments. I say that instead of calling it science fiction, we call it science vision."

REFERENCES

E. Von Daniken, *Presence des extra-terestres; Bibliotheque des grandes enigmes*, 1970.

Wikipedia.org/wiki/Differential_equation.

Printed in the United States
By Bookmasters